Lecture Notes in Computer Science 11382

Commenced Publication in 1973
Founding and Former Series Editors:
Gerhard Goos, Juris Hartmanis, and Jan van Leeuwen

More information about this series at http://www.springer.com/series/7407

Rebeca Marfil · Mariletty Calderón
Fernando Díaz del Río · Pedro Real
Antonio Bandera (Eds.)

Computational Topology in Image Context

7th International Workshop, CTIC 2019
Málaga, Spain, January 24–25, 2019
Proceedings

Springer

Editors
Rebeca Marfil ⓘ
Universidad de Málaga
Málaga, Spain

Mariletty Calderón
Universidad de Málaga
Málaga, Spain

Fernando Díaz del Río ⓘ
University of Seville
Sevilla, Spain

Pedro Real ⓘ
University of Seville
Sevilla, Spain

Antonio Bandera ⓘ
Universidad de Málaga
Málaga, Spain

ISSN 0302-9743   ISSN 1611-3349   (electronic)
Lecture Notes in Computer Science
ISBN 978-3-030-10827-4   ISBN 978-3-030-10828-1   (eBook)
https://doi.org/10.1007/978-3-030-10828-1

Library of Congress Control Number: 2018965613

LNCS Sublibrary: SL1 – Theoretical Computer Science and General Issues

This Springer imprint is published by the registered company Springer Nature Switzerland AG
The registered company address is: Gewerbestrasse 11, 6330 Cham, Switzerland

Preface

This book constitutes the proceedings of the 7th International Workshop on Computational Topology in Image Context, CTIC 2019, held in Málaga, Spain, in January 2019. From the 21 proposals submitted, we selected the present 14 contributions that, after being revised by qualified reviewers, were accepted for presentation. All papers were accepted for oral presentation.

CTIC 2019 was organized by the Engineering of Integrated Systems and the Topological Pattern Analysis and Recognition research groups, from the Universities of Málaga and Seville, respectively. The conference was supported by the University of Málaga through the I Plan Propio de Investigación. We also thank the Higher Technical School of Telecommunications Engineering for hosting the event and providing all the necessary facilities. As with previous editions, CTIC 2019 was also accepted for publication by Springer as an LNCS proceedings volume.

The papers deal with theoretical issues but most of them place the attention on the applicability of concepts and algorithms. These were designed to deal with objects and images, but also with the speech signal. The final application must be, for instance, in the medical domain or in the robotics one.

It has been a great honor for us to count on the participation of two international renowned researchers as invited speakers: Carme Torras, Research Professor at the Spanish Scientific Research Council (CSIC) and head of the Perception and Manipulation group at the Robotics Institute (CSIC-UPC) in Barcelona, and Herbert Edelsbrunner, Professor of Mathematics and Computer Science at the Institute of Science and Technology Austria (IST Austria) in Klosterneuburg.

Finally, we would like to express our gratitude to the Scientific Committee members for their helpful comments, which enabled the authors to improve the quality of their contributions.

January 2019

Rebeca Marfil
Mariletty Calderón
Fernando Díaz del Río
Pedro Real
Antonio Bandera

Organization

CTIC 2019 was organized by the Engineering of Integrated Systems and the Topological Pattern Analysis and Recognition research groups, from the Universities of Málaga and Seville, respectively.

Organizing Committee

Rebeca Marfil	University of Málaga, Spain
Mariletty Calderon	University of Málaga, Spain
Fernando Díaz del Río	University of Seville, Spain
Pedro Real	University of Seville, Spain
Antonio Bandera	University of Málaga, Spain

Scientific Committee

Jacques-Olivier Lachaud
Pascal Lienhardt
Joakim Lindblad
Jeac-Luc Mari
Serge Miguet
Helena Molina-Abril
Marian Mrozek
Mateusz Juda
Nicolas Normandt
Nicolas Passat
Patrizio Frosini
Laurent Funchs
Herbert Edelsbrunner
Pawel Pilarczyk
Sanjoy Pratihar
Pedro Real
Eric Remy
Tristan Roussillon
Ana Romero

Gabriella Sanniti di Baja
Henrik Schulz
Isabelle Sivignon
Natasa Sladoje
Michela Spagnuolo
Robin Strand
Edouard Thiel
Ricardo Uribe Lobello
Antoine Vacavant
José Antonio Vilches
Sophie Viseur
Alexandra Bac
Reneta Barneva
Arindam Biswas
Isabelle Bloch
Srecko Brlek
Didier Coquin
Michel Couprie
Guillaume Damiand

Leila de Floriani
Massimo Ferri
Fabien Feschet
Rocio Gonzalez-Diaz
Aldo Gonzalez-Lorenzo
Isabelle Debled
María J. Jiménez
Bertrand Kerautret
Reinhard Klette
Walter Kropatsch
Guillermo Barcena
Pedro Galindo
Raul Reina Molina
Darian M. Onchis
Vitaliy Kurlin
Akira Nakamura
Nicolai Petkov
María Carmen Escribano

Sponsoring Institution

I Plan Propio de Investigación, Universidad de Málaga, Spain

Contents

Discrete Analog of the Jacobi Set for Vector Fields

A. N. Adilkhanov[1], A. V. Pavlov[2](✉), and I. A. Taimanov[3]

[1] National Laboratory "Astana", Nazarbayev University, Kabanbay Batyr Ave. 53,
Astana 010000, Republic of Kazakhstan
aadilkhanov@nu.edu.kz
[2] North-Eastern Federal University, 677000 Yakutsk, Russia
av.pavlov@s-vfu.ru
[3] Sobolev Institute of Mathematics, Novosibirsk State University,
630090 Novosibirsk, Russia
taimanov@math.nsc.ru

Abstract. The Jacobi set is a useful descriptor of mutual behavior of functions defined on a common domain. We introduce the piecewise linear Jacobi set for general vector fields on simplicial complexes. This definition generalizes the definition of the Jacobi set for gradients of functions introduced by Edelsbrunner and Harer.

Keywords: Jacobi set · Vector fields · Simplicial complex

1 Introduction

In this article we give a construction of a piecewise linear analog of the Jacobi set for vector fields. This set serves as a descriptor of the relation between vector fields defined on a common domain.

For the gradient fields of Morse functions $f_1, \ldots, f_k : D \to \mathbb{R}$, where D is a domain in $\mathbb{R}^N$, or more generally an N-dimensional manifold, the Jacobi set is the subset of D formed by all points at which the gradients of these functions are linearly dependent. This set can be used for extracting useful information about the mutual behavior of multiple functions [1]. As Jacobi sets for a pair of functions on the plane it appears for different reasons in [2] (see also [3]), and in general form it was introduced in [4].

For applications, it is helpful to have a discrete analog of the Jacobi set, and such an analog for functions defined on triangulated complexes was introduced in [4]. In the same article, the problem of extending the proposed methods to general vector fields was posed. We demonstrate how to do that on the example of pairs of vector fields on the plane.

This work was supported by the Ministry of Education and Science of the Republic of Kazakhstan (program 0115PK03029) and Russian Foundation for Basic Research (grant 15-01-01671a).

© Springer Nature Switzerland AG 2019
R. Marfil et al. (Eds.): CTIC 2019, LNCS 11382, pp. 1–11, 2019.
https://doi.org/10.1007/978-3-030-10828-1_1

2   The Piecewise Linear Jacobi Set

We recall the main definitions and results from [4].

A *Morse function* on a compact manifold M is a function $f : M \to \mathbb{R}$ that has only a finite number of critical points, where the matrix of second derivatives is nondegenerate and the function values are distinct from each other. For two Morse functions $f, g : M \to \mathbb{R}$ defined on a compact manifold M, their *Jacobi set* is defined as the set $J(f, g)$ where their gradients are linearly dependent. Equivalently, $J(f, g)$ can be described as the set of all critical points of functions $h_\lambda = f + \lambda g$ and $e_\lambda = \lambda f + g$ for all $\lambda \in \mathbb{R}$. For two generic Morse functions f, g having no common critical points, $J(f, g)$ is a 1-submanifold of M.

In the discrete case of two functions f, g defined on the vertices of a triangulation K of a d-dimensional manifold, we can extend f and g to PL-functions on the entire complex K and view each as a limit of a series of smooth functions. Motivated by this viewpoint, the discrete Jacobi set $\mathbb{J}(f, g)$ is introduced [4] as a 1-dimensional subcomplex of K consisting of edges uv, with multiplicity, along which, in the limit, the critical points of h_λ and e_λ travel as λ varies.

To state the precise definition, we need some notation. Let K be a simplicial complex. The *star* of its simplex σ is the set of all simplices containing σ, and the *link* $\mathrm{Lk}\,\sigma$ consists of all simplices in the closure of the star of σ that are disjoint from σ. Note that $\mathrm{Lk}\,\sigma$ is itself a complex. Let h be a real-valued function on the vertices of a simplicial complex K. For a simplex $\sigma \in K$, define the lower link $\underline{\mathrm{Lk}}\,\sigma$ in K with respect to h to be the portion of $\mathrm{Lk}\,v$ that bounds the set of all simplices in the star of v that have v as the vertex with the maximal value of h.

Consider an edge $uv \in K$. We disregard the edges with $g(u) = g(v)$. Denote by $\lambda^* = \lambda^*(u, v)$ the value of λ that equalizes the values of the linear combination $f + \lambda g$ at both ends of the edge: $f(u) + \lambda^* g(u) = f(v) + \lambda^* g(v)$. Denote this linear combination by h: $h = f + \lambda^* g$. The link of the vertex u is a triangulation of a $(d - 1)$-sphere containing v. The multiplicity of the edge uv is defined as the sum of reduced Betti numbers $\tilde{\beta}_i$ of the lower link $\underline{\mathrm{Lk}}\,uv$ with respect to h. The *piecewise linear Jacobi set* of two functions f, g on K is defined as the one-dimensional subcomplex $\mathbb{J}(f, g)$ of K consisting of all edges having nonzero multiplicity, together with their endpoints.

We now review the general definition for the special cases of 2- and 3-dimensional simplicial complexes.

In two dimensions, the star of an edge uv consists of two 2-simplices neighboring along uv, and $\mathrm{Lk}\,uv$ is just two vertices a, b opposite uv in these 2-simplices (Fig. 1a). Thus, the edge uv belongs to $\mathbb{J}(f, g)$ if and only if the values $h(a)$, $h(b)$ are either both greater or both smaller than the value of $h(u) = h(v)$, where $h = f + \lambda^* g$, and $\lambda^* = (f(u) - f(v))/(g(v) - g(u))$. This condition can be rewritten in terms of function differences in adjacent vertices. For any function $f : K \to \mathbb{R}$ and edge xy, denote $df(xy) = f(y) - f(x)$. Then we can write

$$uv \in \mathbb{J}(f, g) \Leftrightarrow dh(ua) \text{ and } dh(ub) \text{ have the same sign},$$
$$\text{where } h = f + \lambda^* g \text{ and } \lambda^* = df(uv)/df(vu). \tag{1}$$

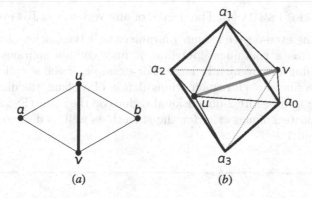

Fig. 1. The link of an edge in a triangulation of (*a*) 2-manifold (*b*) 3-manifold

This condition is actually symmetric in u and v, since $dh(vx) = dh(ux) - dh(uv)$ for any vertex x, and $dh(uv) = 0$.

In a triangulation of a three-dimensional manifold, the link $\mathrm{Lk}\, uv$ of an edge uv is a triangulation of a circle. The multiplicity of an edge uv in the Jacobi set $\mathbb{J}(f, g)$ is equal to the sum of the reduced Betti numbers of the lower link $\underline{\mathrm{Lk}}\, uv$ with respect to h:

$$\tilde{\beta}_{-1} + \tilde{\beta}_0 + \tilde{\beta}_1,$$

where $\tilde{\beta}_{-1}$ is 1 if the lower link is empty, and 0 otherwise; $\tilde{\beta}_0$ is one less than the number of connected components in $\underline{\mathrm{Lk}}\, \underline{uv}$ if this number is positive, and 0 otherwise; $\tilde{\beta}_1$ is 1 if the lower link is the entire circle, and 0 otherwise.

In Fig. 1b, the link of uv is shown in bold lines. This link is a triangulation of a circle. Denote its consecutive vertices by $a_0, a_1, \ldots, a_{k-1}$, and put $a_k = a_0$. As previously, $dh(xy)$ stands for the difference $h(y) - h(x)$. Count the number of times the difference $dh(ua_i)$ changes from negative to positive along the circle:

$$\beta_0 = \#\{i : \ dh(ua_i) < 0 \text{ and } dh(ua_{i+1}) > 0, \ 0 \le i < k\}.$$

Then the multiplicity of the edge uv in $\mathbb{J}(f, g)$ is $|\beta_0 - 1|$. In particular, if the lower link of the edge uv with respect to h consists of just a single component that is not the entire circle, the edge does not belong to the Jacobi set.

The functions $h_\lambda(v) = f + \lambda g$ are linear in λ for any given v. Because of that, any $v \in \mathrm{Lk}\, u$ changes its status as inside/outside the lower link of u with respect to h_λ exactly once as λ grows from $-\infty$ to $+\infty$, namely at $\lambda = \lambda^*(u, v) = df(uv)/dg(vu)$. So in dimension 2, the number of connected components β_0 in the lower link of u is either the same at both extremes, or 0 for one of them and 1 at the other. Obviously, for an edge $uv \notin \mathbb{J}$, passing $\lambda^*(u, v)$ does not change β_0. For $uv \in \mathbb{J}$, passing λ^* either changes β_0 by one or does not change $\beta_0 = 1$ if on either side of λ^* the lower link is all of the link of u. Counting the parity of β_0, we see that the number of edges $uv \in \mathbb{J}$ for a fixed vertex u, i.e. the degree of u in $\mathbb{J}$, must be even. A similar argument, after unfolding each multiple critical point into multiple simple critical points, holds in any dimension:

EVEN-DEGREE LEMMA [4]. The degree of any vertex u in $\mathbb{J}$ is even.

Although the even-degree lemma guarantees that the discrete Jacobi set can be represented as a continuous polyline, it may contain spurious cycles and zigzags, becoming inconveniently large. For example, if the simplicial complex in question is a fine enough regular triangulation of a plane, the discrete Jacobi set may appear to fill entire 2-dimensional regions on the plane (Fig. 2). A variety of simplification techniques exist for the smooth as well as discrete versions of the Jacobi set [6,7].

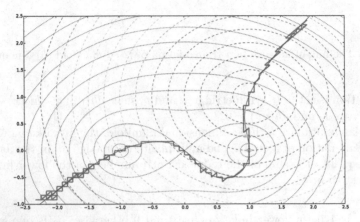

Fig. 2. Zigzags in a simplicial Jacobi set (red) for functions $f = ((x-1)^2 + y^2)((x+1)^2 + y^2)$ (dotted level lines) and $g = (x-1)^2 + (y-1)^2$ (dashed level lines). The triangulation is obtained from a square grid with step $\frac{1}{6}$. The continuous black line is the smooth Jacobi set. (Color figure online)

3   The Piecewise Linear Jacobi Set for Vector Fields

The main idea behind our definition is as follows. The gradient

$$df = \left(\frac{\partial f}{\partial x^1}, \ldots, \frac{\partial f}{\partial x^N} \right)$$

of a function $f : D \to \mathbb{R}$ is in fact a 1-form which is a linear form on vector fields. Indeed, its value for a vector field X is the derivative of f in the direction of X:

$$D_X f = X^i \frac{\partial f}{\partial x^i},$$

where we assume the summation over the repeated index. To obtain the gradient vector field we have to raise the index by using some non-degenerate quadratic form g^{ik} (usually the inverse of the metric tensor g_{ik}):

$$(\nabla f)^i = g^{ik} \frac{\partial f}{\partial x^k},$$

where again we assume the summation over the repeated index k. The Euclidean metric is given by the tensor

$$g_{ik} = g^{ik} = \begin{cases} 1 & \text{for } i = k \\ 0 & \text{otherwise,} \end{cases}$$

the gradient of the function and the gradient vector field look the same, but in general coordinates their numerical expressions are different. We refer for detailed discussion, for instance, to [5].

Since the lowering of the index (the convolution)

$$X^i \rightarrow Y_k = g_{ik} X^i$$

maps linearly dependent vector fields into linearly dependent 1-forms, it is enough to define the Jacobi sets for 1-forms.

For a triangulated complex K, 1-forms Y are linear functions on oriented 1-chains, i.e., on oriented edges:

$$Y(uv) \in \mathbb{R} \text{ where } uv \text{ is an oriented edge in } K.$$

We interpret the Edelsbrunner–Harer definition of the Jacobi set of two gradient vector fields as the definition of the Jacobi set of two 1-forms that are coboundaries of linear functions on the vertices of K:

$$Y(uv) = df(uv) = f(v) - f(u).$$

For a triangulation of a smooth manifold K and a smooth function $f : K \rightarrow \mathbb{R}$ the discretization of its gradient (covector) field is exactly given by the formula above where f is evaluated in the vertices of the triangulation.

Given a smooth 1-form ω on a triangulated manifold, we have to construct a 1-form on oriented edges. The most natural way is to consider an edge as an oriented path and take an integral of ω over the path:

$$Y(uv) = \int_u^v \omega.$$

For a smooth gradient field df in Euclidean space we get

$$Y(uv) = \int_v^u df = f(v) - f(u).$$

A non-gradient vector field corresponds to a non-closed 1-form. Circular integrals of such a form may not vanish, so generally, it is not true that $Y(uw) = Y(uv) + Y(vw)$.

Let K be a simplicial complex that is a triangulation of a d-dimensional manifold, and F, G be discrete 1-forms given by their values on all oriented edges uv of K:

$$F(uv) = -F(vu), \quad G(uv) = -G(vu).$$

Denote by H_λ the linear combination $F + \lambda G$. For each edge uv with $G(uv) \neq 0$, as previously, denote by λ^* the value of the coefficient that makes this linear combination vanish along uv:

$$\lambda^* = \frac{F(uv)}{G(vu)}, \quad H_{\lambda^*}(uv) = 0.$$

For a vertex w in Lk uv define the average of the values of the form H_{λ^*} on the edges connecting u and v to w:

$$h(w) = \frac{1}{2}(H_{\lambda^*}(uw) + H_{\lambda^*}(vw)).$$

Multiplicity of an edge uv is defined as the sum of the reduced Betti numbers of the lower link of uv with respect to h, and we define the *Jacobi set* $\mathbb{J}(F, G)$ *of two discrete 1-forms F and G* as the one-dimensional subcomplex of K consisting of all edges having nonzero multiplicity, together with their endpoints.

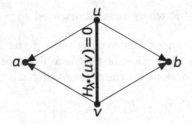

Fig. 3. The edge test for 1-forms on a 2-dimensional complex

In two dimensions ($d = 2$), this definition means that the Jacobi set of F and G consists of all edges uv for which the average of the values of $H_{\lambda^*} = F + \lambda^* G$ along ua and va has the same sign as the average of its values along ub and vb, where a and b are the two points of the link of uv (Fig. 3).

$$uv \in \mathbb{J}(F, G) \Leftrightarrow (H_{\lambda^*}(ua) + H_{\lambda^*}(va))(H_{\lambda^*}(ub) + H_{\lambda^*}(vb)) > 0. \qquad (2)$$

Note that, as was the case for the condition (1), this condition is also symmetric in u, v. It is symmetric in F and G as well when all values of the forms F, G on the edges are nonzero.

However, the even degree lemma no longer holds for nongradient 1-forms. This is illustrated below for the approximation of the Jacobi set for two smooth 1-forms on the plane (Fig. 4). The smooth Jacobi set is the set of points where the forms are linearly dependent, and is shown with continuous green lines, while the piecewise linear Jacobi set for the triangulation of a square grid with step size $h = 0.1$ is shown in red.

Still, as can be seen in Table 1, with the refinement of the grid the approximation converges to the smooth Jacobi set.

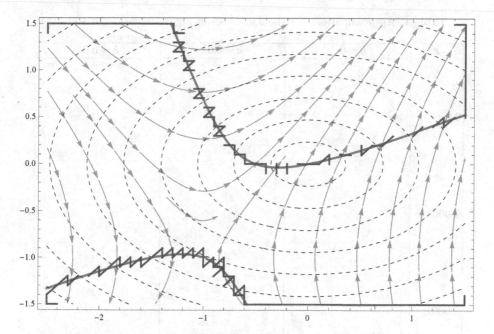

Fig. 4. Continuous Jacobi set (green line) and its piecewise linear approximation for the forms $F(x,y) = (y + 1)dx + 2(x + 1)dy$ and $G(x,y) = (2x - 3y)dx + (2x + 3y)dy$, grid step size $h = 0.1$ (Color figure online)

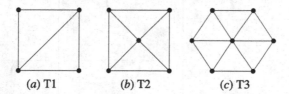

(a) T1   (b) T2   (c) T3

Fig. 5. Plane triangulations

In applications, a vector field X is usually given by its coordinates on a plane grid. A reasonable approximation for the integrals of the corresponding 1-form Y on the edges are scalar products of the mean value of the vector field on the edge with the edge vector itself:

$$Y(uv) = \frac{1}{2}(X_v + X_u, uv).$$

We have also tested our definition for three different regular triangulations on the plane, shown in Fig. 5: the diagonal grid T_1 (invariant with respect to rotations by π), crossed T_2 (invariant with respect to rotations by $\pi/4$), and hexagonal T_3 (invariant with respect to rotations by $\pi/6$).

Table 1. Approximations for the Jacobi set in Fig. 7 for various h and ε

Results of these calculations are shown in Fig. 6. As was the case for the Jacobi sets of Morse functions, the approximations differ, with no clear winner.

For better connectivity of the produced approximation, the edge test (2) can be modified to include cases where the absolute value of at least one of the factors is smaller than some threshold value ε:

$$\mathbb{J}(F, G) = \{uv \in K \mid (H_{\lambda*}(ua) + H_{\lambda*}(va))(H_{\lambda*}(ub) + H_{\lambda*}(vb)) > 0.$$
$$\text{or } |H_{\lambda*}(ua) + H_{\lambda*}(va)| < \varepsilon \quad \text{or } |H_{\lambda*}(ub) + H_{\lambda*}(vb)| < \varepsilon\} \quad (3)$$

This will improve the connectivity at the cost of thickening the Jacobi set.

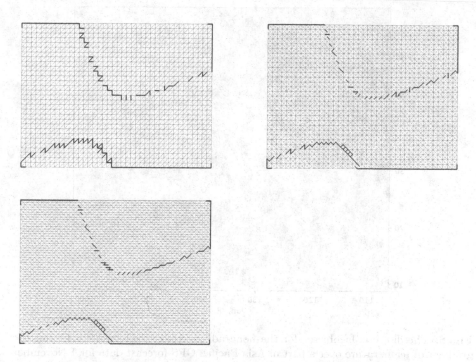

Fig. 6. Piecewise linear Jacobi set of $F(x,y) = (y+1)dx + 2(x+1)dy$ and $G(x,y) = (2x - 3y)dx + (2x + 3y)dy$ for various triangulations on the plane

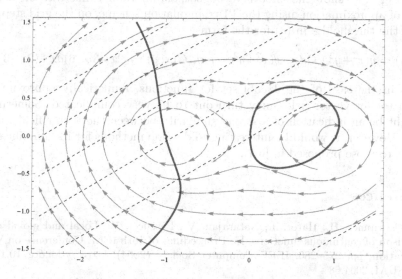

Fig. 7. The fields f, g and the smooth Jacobi set $J(F,G)$ for the corresponding forms $F(x,y) = y(x^2 + y^2 + 1)dx - x(x^2 + y^2 - 1)dy$, $G(x,y) = (2x - 3y - 6)dx + (2x - 3y)dy$

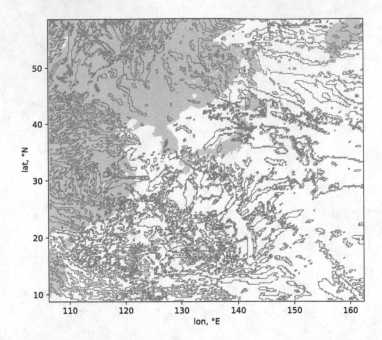

Fig. 8. The discrete Jacobi set for the nongradient wind speed vector field and the gradient of temperature over a part of Asia Pacific. GFS forecast data for 1 November 2018 used [8].

In Fig. 7, we show the smooth Jacobi set, and in Table 1 illustrate the dependence of approximation, using the T1 triangulation scheme, on the grid step size h and the threshold ε in (3) for the forms

$$F(x,y) = y(x^2+y^2+1)dx - x(x^2+y^2-1)dy, \; G(x,y) = (2x-3y-6)dx+(2x-3y)dy.$$

As in the case of the Jacobi set for functions, numerically approximated Jacobi set for vector fields may turn out to be very complicated. Sometimes, it might be an indication of a strong similarity between the vector fields, as in Fig. 8. However, it would be interesting to develop methods for its simplification similar to those proposed in [6,7].

References

1. Edelsbrunner, H., Harer, J., Natarajan, V., Pascucci, V.: Local and global comparison of continuous functions. In: Proceedings of 16th IEEE Conference on Visualization, pp. 275–280. IEEE Computer Society (2004). https://doi.org/10.1109/VISUAL.2004.68
2. Wolpert, N.: An exact and efficient approach for computing a cell in an arrangement of quadrics. Ph.D. thesis, Universität des Saarlandes (2002)

3. Wolpert, N.: Jacobi curves: computing the exact topology of arrangements of non-singular algebraic curves. In: Di Battista, G., Zwick, U. (eds.) ESA 2003. LNCS, vol. 2832, pp. 532–543. Springer, Heidelberg (2003). https://doi.org/10.1007/978-3-540-39658-1_49

4. Edelsbrunner, H., Harer, J.: Jacobi sets of multiple morse functions. In: Cucker, F., DeVore, R., Olver, P., Süli, E. (eds.) Foundations of Computational Mathematics, Minneapolis 2002. London Mathematical Society Lecture Note Series, pp. 37–57. Cambridge University Press, Cambridge (2004). https://doi.org/10.1017/CBO9781139106962.003

5. Novikov, S., Taimanov, I.: Modern Geometric Structures and Fields. American Mathematical Society, Providence (2006)

6. Natarajan, N.S.: Simplification of jacobi sets. In: Pascucci, V., Tricoche, X., Hagen, H., Tierny, J. (eds.) Topological Methods in Data Analysis and Visualization: Theory, Algorithms, and Applications. MATHVISUAL, pp. 91–102. Springer, Heidelberg (2011). https://doi.org/10.1007/978-3-642-15014-2_8

7. Bhatia, H., Wang, B., Norgard, G., Pascucci, V., Bremer, P.T.: Local, smooth, and consistent jacobi set simplification. Comput. Geom.: Theory Appl. **48**(4), 311–332 (2015). https://doi.org/10.1016/j.comgeo.2014.10.009

8. NOAA Operational Model Archive and Distribution System. Data Transfer: NCEP GFS Forecasts (0.25 degree grid). http://nomads.ncep.noaa.gov/cgi-bin/filter_gfs_0p25_1hr.pl?dir=%2Fgfs.2018110100

Computing and Reducing Slope Complexes

Walter G. Kropatsch[1]([✉]), Rocio M. Casablanca[2], Darshan Batavia[1],
and Rocio Gonzalez-Diaz[2]

[1] Pattern Recognition and Image Processing Group 193/03,
TU Wien, Vienna, Austria
{krw,darshan}@prip.tuwien.ac.at
[2] University of Seville, Seville, Spain
{rociomc,rogodi}@us.es

Abstract. In this paper we provide a new characterization of cell decomposition (called slope complex) of a given 2-dimensional continuous surface. Each patch (cell) in the decomposition must satisfy that there exists a monotonic path for any two points in the cell. We prove that any triangulation of such surface is a slope complex and explain how to obtain new slope complexes with a smaller number of slope regions decomposing the surface. We give the minimal number of slope regions by counting certain bounding edges of a triangulation of the surface obtained from its critical points.

1 Introduction

Discrete representations of surfaces in 2.5D like images or digital terrain models are discretizations of 2-dimensional (2D) continuous surfaces. Important properties of such surfaces are their critical points: local minima, maxima and saddle points. These points can be connected by monotonic paths that either go up or go down. These paths delineate surface patches that can be characterized by the property that every pair of points inside such a patch can be connected by a monotonic path: slope regions. Slope regions may be seen as "filling the space between the critical points of the surface" [8]. A planar triangle is an example of a slope region and any triangular mesh subdivides the surface into a set of slope regions. Critical points can appear in many different configurations. Also the subdivision into slope regions may take different arrangements.

In this paper, we explain how to create and reduce slope complexes (decomposition of the given 2D continuous surfaces in slope regions), and we also address the question whether there is a minimal number of slope regions that completely fill the surface between a given set of critical points.

Similar considerations have been published by Edelsbrunner et al. [3–5] with the intention to construct a hierarchy of increasingly coarse Morse complexes. The concept of *'integral line'* (defined in [3]) has great similarity to the monotonic paths of our approach although we may have less geometric constraints.

© Springer Nature Switzerland AG 2019
R. Marfil et al. (Eds.): CTIC 2019, LNCS 11382, pp. 12–25, 2019.
https://doi.org/10.1007/978-3-030-10828-1_2

Specifically, our definition does not need smooth surfaces and monotonic paths do not necessarily follow the steepest slope.

The paper is organized as follows: First we give some basic definitions. We then introduce slope complexes as abstract cellular complexes built by slope regions. We then study the properties of the boundaries of the slope regions and identify particular slope regions, named simple and non-simple triangles, that serve us to construct slope complexes composed only by triangles. We then enumerate the conditions to merge slope regions, the basic process to reduce the number of slopes without affecting the bounding critical points. Finally, we compute the minimum number of slope regions needed to cover the surface given its critical points.

2   Preliminaries

Let us introduce the terminology and main definitions which will be used throughout this paper.

Given a continuous function $g : \Re^2 \mapsto \Re$, a 2-dimensional (2D) continuous surface $S = \{(x, y, z) \in \Re^3 : z = g(x, y)\}$ can be defined. Given a point $p = (x, y, z) \in S$, we sometime denote $g(x, y)$ by $g(p)$ by abuse of notation.

Definition 1 (local neighborhood). *Let $p \in \Re^k$ where $k = 1, 2, \ldots$ and $r \in \Re$, with $r > 0$. The local neighborhood of p is a k-dimensional open ball of radius r and center p, denoted by $B_k(p, r)$, that is the set of points $q \in \Re^k$ such that $d(p, q) < r$.*

Definition 2 (1-extrema). *Let $a, b \in \Re$, with $a < b$. Let $\gamma : [a, b] \to \Re^2$ be a continuous curve and $t \in [a, b]$. If there exists $\epsilon > 0$ such that $g(\gamma(t)) \leq g(\gamma(s))$, for every $s \in B_1(t, \epsilon) \cap [a, b]$, then $\gamma(t)$ is a 1-minimum. Similarly, if there exists $\epsilon > 0$ such that $g(\gamma(t)) \geq g(\gamma(s))$, for every $s \in B_1(t, \epsilon) \cap [a, b]$, then $\gamma(t)$ is a 1-maximum. Finally, $\gamma(t)$ is a 1-extremum if it is a 1-maximum or a 1-minimum.*

Definition 3 (monotonic path). *Let $a, b \in \Re$, with $a < b$. A monotonic path $\pi : [a, b] \to \Re^2$ between $p = \gamma(a)$ and $q = \gamma(b)$ is a continuous curve satisfying that there is no $t \in (a, b)$ such that $\gamma(t)$ is a 1-extrema.*

A level curve is a particular case of monotonic path.

Definition 4 (level curve). *Let $a, b \in \Re$, with $a < b$. A level curve $\gamma : [a, b] \to \Re^2$ is a continuous function such that there exists a constant $c \in \Re$ where $g(\gamma(t)) = c$, for all $t \in [a, b]$.*

A monotonic path is either non increasing or non decreasing and, then, it is always bounded by a 1-maximum and a 1-minimum. This allows us to provide monotonic curves (excluding level curves) with a natural orientation (in our illustrations: an arrow from point a to point b means an edge with endpoints a and b such that $g(a) > g(b)$.

Different types of points in S can be described depending on their 2-dimensional local neighborhood.

Definition 5. *Let p be a point of $\Re^2$. Three categories for p can be distinguished:*

- *For a 2-minimum point p there exists $r > 0$ such that $g(p) < g(q)$, for every $q \in B_2(p, r)$.*
- *For a 2-maximum point p there exists $r > 0$ such that $g(p) > g(q)$, for every $q \in B_2(p, r)$.*
- *A point p is a saddle point if for all $r > 0$ there are two points that cannot be connected by a monotonic path in $B_2(p, r)$.*

A 2-extremum is either a 2-minimum or a 2-maximum and a critical point is either a 2-extremum or a saddle point. Critical points are also referred to and denoted as follows: 2-min $\ominus$, 2-max $\oplus$ and saddle $\otimes$.

Observe that two 2-maxima (resp. 2-minima) cannot be connected by a monotonic path. Equivalently, two 1-maxima (resp. 1-minima) cannot be connected by a monotonic path (except for level curves).

Remark 1. In this paper, we exclude plateaus (connected component of points with the same g-value) from the considered surfaces that emerge from the expansion of critical points.

3   Slope Complexes

Roughly speaking, a finite regular CW complex [6] can be seen as a partition, in basic building blocks called cells, of a given topological space X. More concretely, for each k-dimensional cell (k-cell) c in the partition of X, there exists a homeomorphism f (attaching map) from the k-dimensional closed ball to X such that the restriction of f to the interior of the closed ball is a homeomorphism onto the cell c, and the image of the boundary of the closed ball is a homeomorphism onto the union of a finite number of cells of the partition, each having dimension less than k. The closed k-cell $\bar{c}$ is the image of such homeomorphism f.

The CW complexes considered in this paper will be cell decomposition, denoted by $K[S]$, of the 2D continuous surface S, obtained from a continuous function $g : \Re^2 \mapsto \Re$, satisfying that all critical points of the surface are 0-cells.

Observe that only 0-, 1- and 2-cells are permitted in $K[S]$. From now on, we use equivalently the notions vertex, edge, and region as 0-cell, 1-cell and 2-cell respectively similar to [3]. Finally, observe that the boundary of a 2-cell is a continuous closed curve.

Let us introduce the main concept of our paper, *slope regions*, which are different to the regions defined in [3].

Definition 6 (slope region). *Let $K[S]$ be a cell decomposition of a 2D continuous surface S. A slope region R is a 2-cell in $K[S]$ with the constraint that all pairs of points in $\overline{R}$ are connected by a monotonic path inside $\overline{R}$, where $\overline{R}$ is the closure of R (that is, R together with its boundary).*

Definition 7 (slope complex). *A cell decomposition $K[S]$ of a 2D continuous surface S is a slope complex if all its 2-cells are slope regions.*

Now we describe the boundary of any slope region.

Lemma 1. *The boundary of a slope region is composed by either a level curve or two monotonic paths connecting a 1-maximum and a 1-minimum.*

Proof. The boundary of the slope region is a continuous closed curve $\gamma : [a, b] \to \Re^2$, since a slope region is a 2-cell of $K[S]$. This curve can be a level curve, which is a trivial monotonic path. Alternatively, suppose that the g-values (i.e. the values of g) along the boundary vary. Consider the values of $g(\gamma(t))$, with $t \in [a, b]$. Reasoning now by contradiction, assume that $\gamma(t)$ have two 1-maxima and two 1-minima. Since $\gamma(t)$ is a continuous curve, observe that 1-minima and 1-maxima alternate along $\gamma(t)$. By definition of a slope region, the two 1-maxima are connected by a monotonic path, denoted by π_{max}, inside the slope region. All points along π_{max} have g-values between the two 1-maxima, that is, not below the smallest 1-maximum. The path π_{max} splits the slope region into two or more sub-regions. The two 1-minima appear in two different sub-regions and they have smaller g-values than the smallest maximum. By the definition of slope region, there is also a path between the two 1-minima, denoted by π_{min}, which cross π_{max} because the extrema are alternating along $\gamma(t)$. Let us see that π_{min} cannot be monotonic. Let $p = \pi_{max} \cap \pi_{min}$. Notice that $g(p)$ is a value greater than or equal to the smallest 1-maximum and the smallest 1-maximum is greater than all the possible g-values between the two 1-minima. It means that the g-values in π_{min} first increase from one 1-minimum to $g(p)$ and then the g-values decrease from $g(p)$ to the other 1-minimum. Hence, π_{min} is not a monotonic path. Consequently there exists only one local 1-maximum and one local 1-minimum along the boundary of a slope region. □

The boundary of a slope region R can also be folded such that a part of the boundary lies "inside" the region R. When following the boundary such parts are traversed twice.

Definition 8 (inner and outer boundary point). *Let R be a slope region bounded by a continuous closed curve $\gamma : [a, b] \to \Re^2$. Let $p \in \Re^2$ be a point for which there exists $t \in [a, b]$ such that $\gamma(t) = p$. The point p is an inner point of R if there exists $r > 0$ such that $B_2(p, r) \setminus \Gamma \subseteq R$, being $\Gamma = \{\gamma(t) : t \in [a, b]\}$. The outer boundary of R is the set of points of Γ that are not inner boundary points.*

Observe that the outer boundary of R is a simple continuous closed curve.

The following result characterizes the critical points on the inner boundary of a slope region. They can be 2-extrema but never saddle points.

Lemma 2. *The boundary of a slope region R may contain as inner boundary points: a 2-maximum, a 2-minimum, or both simultaneously, but never a saddle point.*

Proof. Neither two 2-maxima nor two 2-minima can be connected by a monotonic path, then the first part of the statement is trivial. Now, let us see that a saddle point cannot be an inner boundary point of a slope region. A saddle point is characterized by its local neighborhood. By contradiction, assume that R is a slope region with an inner boundary point x being a saddle point. Consider a small enough r such that $B_2(x,r) \subseteq R$, then, by definition of saddle point, there exists p and q in $B_2(x,r)$ such that there is no monotonic path between them, which is a contradiction with the definition of slope region.   □

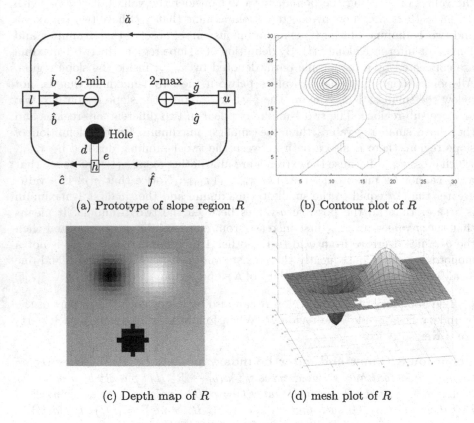

(a) Prototype of slope region R (b) Contour plot of R

(c) Depth map of R (d) mesh plot of R

Fig. 1. Prototype of a slope region R.

We finish this section by giving a prototype of a general slope region R. Figure 1a shows all the components that R can have. We use following notation in describing a traversal of a complete boundary: curves with arrows indicate descending orientation in g-values, the boundary segments in which the boundary curve is subdivided are denoted with characters a, b, c, d, e, f and g in counter clockwise orientation around the boundary of R with notations ˆ and ˇ, where for example $\hat{f}$ denotes that we follow the boundary uphill and $\breve{a}$ denotes that we follow the boundary downhill (see Fig. 1d for the intuition of descending, ascending, downhill and uphill).

Following the *outer boundary* we encounter following monotonic paths: $\breve{a}, \hat{c}, \hat{f}$ connecting the up-most (highest g-value) point u with the lowest point l. While the *inner boundary* includes $\breve{b}, \hat{b}, d, e, \hat{g}, \breve{g}$. Boundary segments d and e are *level curves* triggered by the hole $\bullet$ and has no orientation. All inner boundaries are single monotonic paths connecting the outer boundary to the inner boundary points of R which are traversed twice and return to the same outer boundary point where they started. A complete traversal of the boundary of this slope region R is

$$(u, \breve{a}, l, \breve{b}, \ominus, \hat{b}, l, \hat{c}, h, d, \bullet, e, h, \hat{f}, u, \hat{g}, \oplus, \breve{g}).$$

Furthermore, R may encapsulate one or more other slope regions which for R appear as holes. See Fig. 1b in which the boundary of the hole is spotted in the contour plot of a grayscale image (Fig. 1c) with similar configuration as prototype. Figure 1d is the mesh plot, where the height corresponds directly to the intensity (g-value) of the image. The boundary of these holes and its connection d and e to the point h on the outer boundary of R should all be level curves.

The other possible cases of the slope region R may exclude any one of the 2-extrema $\ominus, \oplus$ or the holes $\bullet$. The absence of $\ominus$ will infer that l becomes the 1-minimum of R. Similarly in absence of $\oplus$, u becomes the 1-maximum of R.

The formal proof that this general prototype is always a slope region is left to future work. The idea of the prove is as follows[1]: Take two points p and q in R, then $g^{-1}(g(p))$ and $g^{-1}(g(q))$ are level curves inside R. These level curves: either (1) cross the connection between the 2-max the 2-min or one hole to the outer boundary; or (2) connect the two monotonic paths of the outer boundary. Then it is possible to construct a monotonic path from p to q following the level curves inside R and the monotonic paths in the boundary of R.

4   Creating Slope Complexes

In this section we prove that a triangulation of a 2D continuous surface is a slope complex.

A simple triangle is defined as a simply-connected planar region in $\Re^3$ (i.e. a connected region without holes and embedded in a plane not parallel to xz- or yz-plane in $\Re^3$) bounded by three edges (not necessarily straight line segments) connecting three vertices a, b and c. Figure 2 shows the different types of triangles that exist. We distinguish between simple triangles (Fig. 2a) and non-simple triangles (Fig. 2b and c) where two of the three vertices coincide. Observe that the edge from a to b is counted twice in Fig. 2b.

Lemma 3. *Simple and non-simple triangles are slope regions.*

Proof. Since any (simple or non-simple) triangle T is simply-connected and embedded in a plane, there is a monotonic path connecting any two points in T. $\qquad\square$

---

[1] For "gray value" z, $g^{-1}(z) = \{p \in \Re^2 \mid g(p) = z\}$ is the level set of gray value z.

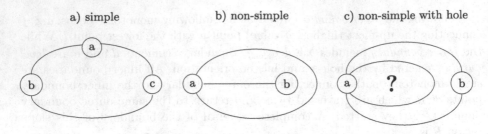

Fig. 2. Simple and non-simple triangles.

As immediate consequence, next result follows.

Corollary 1. *A triangular mesh is a slope complex.*

Let us explore the triangular slope regions. First, consider a simple triangle abc in $\Re^3$ with edges being monotonic paths. Assume $g(a) \geq g(b) \geq g(c)$, then we have the three following cases:

1. If $g(a) = g(b) = g(c)$, the triangle is a plateau (i.e., all points of the triangle having same g-value).
2. If $g(a) = g(b)$ and $g(b) > g(c)$ then, the triangle have one edge with constant g-value, vertices a and b are 1-maxima and vertex c is a 1-minimum. If $g(a) > g(b)$ and $g(b) = g(c)$ we can follow a similar reasoning than above.
3. If $g(a) > g(b) > g(c)$ (Fig. 2a), then, vertex a is a 1-maximum and vertex c is a 1-minimum along the boundary of the triangle and vertex b is in-between. Vertex a does not need to be a 2-maximum since other neighbors outside the triangle can be larger. Analogously, vertex c does not need to be a 2-minimum. Vertex b cannot be a 1-maximum or a 1-minimum. If we restrict our vertices to 2-dimensional critical points then b must be a saddle point. We discuss these cases below (Sect. 6.1).

Second, in the non-simple case, the outer boundary (self-loop attached to vertex a) is a level curve surrounding vertex b. Since $g(b) \neq g(a)$ vertex b is an extremum surrounded by level curves with g-values between $g(a)$ and $g(b)$. The edge connecting vertices a and b is a monotonic path. Any point p inside the non-simple triangle must satisfy that $g(p) \in [\min(g(a), g(b)), \max(g(a), g(b))]$ and can be connected to the edge (a, b) following a level curve.

The third type of triangle (Fig. 2c) contains a self-loop attached to vertex a inside the triangle as a level curve. Observe that a sub-complex bounded by the level curve exist but is not part of the triangle, seen from the triangle it is a hole. In this case $g(a) \neq g(b)$ are the 1-extrema of the triangle and the connections between vertices a and b are monotonic paths. The points in the loop has the same g-value as $g(a)$ and any point of the loop as well as inside the triangle can be connected to vertex b by a monotonic path.

5   Merging of Slope Complexes

In this section we establish conditions to obtain a new slope region as a result of merging two slope regions. The final aim is to start from a slope complex (which can be, for example, a triangular mesh) and produce a new slope complex, homeomorphic to the former, obtained by merging slope regions, decreasing, in this way, the initial number of slope regions of the complex.

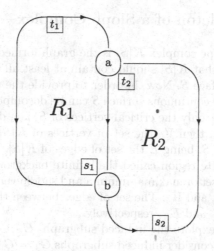

Fig. 3. Merging two slope regions R_1 and R_2

Let R_1 and R_2 be two slope regions sharing a common edge (a, b). Since R_1 and R_2 are slope regions then the edge (a, b) is a level curve or a monotonic path. Without loss of generality, Suppose $g(a) \geq g(b)$. Let t_1, s_1, t_2, s_2 denote the 1-maxima and 1-minima of R_1 and R_2 respectively. Then, the two slope regions R_1 and R_2 can be merged into a new slope region U, by removing the common edge (a, b) if the following properties are satisfied:

1. The common edge is not a self-loop, that is, vertex a is different to vertex b.
2. One of the two 1-minima is on the common boundary.
3. One of the two 1-maxima is on the common boundary.

Now, let us see that there exists a monotonic path between two any points of U:

- Any pair of points in R_1 are connected by a monotonic path because R_1 is a slope region. The same is true for R_2.
- Any pair of points $r_1 \in R_1$ and $r_2 \in R_2$ with $g(a) \geq g(r_1), g(r_2) \geq g(b)$ can be connected by a level curve to points on the common edge (a, b) which is a level curve or a monotonic path.

- We now consider points outside the range of $[g(b), g(a)]$. Let us assume that $g(t_1) > g(a)$ (see Fig. 3). Then, clearly, point t_1 is not on the common edge (a, b). By requirement 3, point t_2 must be on the common edge (a, b). Hence for all points $r_2 \in R_2$, $g(r_2)$ is not higher than $g(t_2)$ and then any point $r_2 \in R_2$ can be connected by a monotonic path to point $r_1 \in R_1$ for $g(r_1) > g(t_2)$. The same holds for the case of $g(t_2) > g(a)$ with the roles of R_1 and R_2 interchanged. Analogously we argue for the 1-minima s_1 and s_2.

6   On the 1-Skeleton of a Slope Complex

The 1-skeleton of a slope complex $K[S]$ is the graph formed by the vertices and edges of $K[S]$. Recall that $K[S]$ should contain at least all the critical points of the 2D continuous surface S. Now, in order to provide the minimum number of slope regions that a 2D continuous surface S can be decomposed, we assume that the vertices of $K[S]$ are only the critical vertices of S, i.e., denote by $G = (V, E)$ the 1-skeleton of $K[S]$, then V (the set of vertices of $K[S]$) coincides with the set of critical points of S, being E the set of edges of $K[S]$. We also assume that there is only one infinite region called the infinite background. Finally, observe that G is planar. The set of maxima, minima and saddle vertices will be denoted respectively by $V_\oplus$, $V_\ominus$ and $V_\otimes$. The set of edges between the set of vertices will be denoted by $E_\oplus$, $E_\ominus$ and $E_\otimes$, respectively.

In Subsect. 6.1 we explore the induced subgraphs $G_\otimes \subset G$ of saddle vertices and in Subsect. 6.2 we consider induced subgraphs $G_\pm = G \backslash G_\otimes \subseteq G$ of extrema only.

6.1   Forest of Saddles

Adding the constraints given in [3] to the 2D continuous surface S, we can assume that saddle points are vertices of $G = (V, E)$ of degree 4, since saddle points with higher degree can be unfolded into a set of connected saddle points of degree 4 (see [3]). Besides, vertices can be characterized by edges incident to it and their respective orientation: (1) all the edges incident to a saddle point have alternating directions; (2) the in-degree of a maximum is 0; and (3) the out-degree of a minimum is 0.

Remark 2. Every extremum on the boundary of any slope region with more than two vertices is adjacent to a saddle.

By Remark 3, saddle points can be connected but cannot form cycles.

Remark 3. Any connected configuration of saddles $G_\otimes = (V_\otimes, E_\otimes) \subset G$ is acyclic, i.e., they form tree structures.

Therefore the saddle points form a forest.

Lemma 4. *Let $G_\otimes = (V_\otimes, E_\otimes)$ be the subgraph of G induced by the set of vertices $V_\otimes$ and let T_i, $i = 1, \ldots n$, be the n maximal connected components of $G_\otimes$ which are trees by Remark 3. Let $V_{\otimes,i}, i = 1, \ldots n$, be the set of vertices of T_i. Let us focus on the edges in $(V_{\otimes,i} \times V) \cap E$ of the ith connected component of $G_\otimes$ and ignore if extrema at the leafs are shared by several trees.*

1. *If the tree T_i contains $|V_{\otimes,i}| = k_i$ connected saddle points then there are $2(k_i + 1)$ pending edges (i.e., edges incident to leafs). The end points of these pending edges are extrema determined by the orientation of the edges. That is, if (s, v) is one of these pending edges then $(s, v) \in V_{\otimes,i} \times V_\ominus$ or $(v, s) \in V_\oplus \times V_{\otimes,i}$.*
2. *Furthermore the pending edges have alternating orientations leading to an alteration of $k_i + 1$ minima and $k_i + 1$ maxima as leafs when moving around the tree.*
3. *The extrema of a connected component of saddle points are connected by monotonic paths formed by the oriented edges of the internal nodes of the tree.*

Proof. We prove the three properties separately.

1. Each saddle point of T_i generates four edges in G. Besides, the k_i saddle points are connected by $k_i - 1$ edges in the tree. Since each edge is counted twice from each end point we have a total of $4k_i - 2(k_i - 1) = 2k_i + 2$ pending edges in each connected component T_i. As end points of the pending edges only extrema are available since all saddle points have been collected in the forest and T_i is maximal.
2. All the edges of a saddle point have alternating directions in clockwise and counter-clockwise orientation. If two saddle points are connected then adjacent leafs are connected by a monotonic path. Since the connected component of saddle points is supposed to be maximal no saddle point can appear as a leaf. Hence the target of an outgoing edge is a minimum and the start of an outgoing pending edge is a maximum.
3. A connected component T_i consists of saddle points inside the tree and of extrema at the leafs. Starting with a maximum we follow the tree downwards keeping the tree always on the same (e.g. left-hand) side. This monotonic path ends in the adjacent minimum. Similarly we can start at a minimum and follow the tree up-wards keeping the tree on the same side. We find the next maximum as end point of a monotonic path.

$\square$

Figure 4a illustrates the main content of the above lemma by an example.

Remark 4. Each connected component T_i is connected to $|V_{\otimes,i}| + 1$ maxima and to the same amount of minima in G. Ignoring the sharing of extrema between different components, altogether the saddle points are connected to $|V_\otimes| + n$ maxima and to $|V_\otimes| + n$ minima in G.

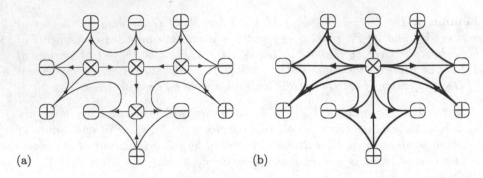

(a)   (b)

Fig. 4. Folding a configuration with four saddles.

Slope regions are generated by connecting the alternating extrema in the order of the leafs of each saddle tree. Observe that this is not necessarily the smallest number of slope regions decomposing the 2D continuous surface S and extrema may not be connected originally.

Remark 5. This new graph is also denoted by $G = (V, E)$ and its faces generate a slope complex $K[S]$ which is again a cell decomposition of S in slope regions.

6.2   Graph of Extrema

Here we consider the subgraph $G_\pm = (V_\pm, E_\pm)$ of G obtained after the removal of all saddles. Then, $V_\pm = V_\oplus \cup V_\ominus$ and $E_\pm = (V_\oplus \times V_\ominus) \cap E$.

These graphs of extrema have another interesting property: they form alternating sequences of maxima and minima connecting an extremum to itself along a closed level curve. In contrast to connected saddle components minimax-sequences can form cycles of lengths 2 (double edge), 4 (non-well-formed), 6 and so on.

Cycles of lengths 2 surround slope regions. The regions surrounded by longer cycles do not form slope regions since they contain more than one minimum and more than one maximum   by Lemma 1.

The cycle of length 4 corresponds to the well-known checkerboard pattern, the non-well-composed configurations [7]. It "hides" a saddle point inside the cycle (compare with [2]). This is true for all longer cycles of extrema, a cycle of $2n$ length needs $n-1$ saddle vertices to subdivide the interior region completely into slope regions. The insertion of one additional (saddle) vertex in such non-well-composed configuration with successive triangulation produces a set of slope regions covering the previous non-slope region [2, Sect. 3.1].

We have seen in Lemma 1 that any slope region with n critical points have a 1-maximum and a 1-minimum in its boundary. Consequently the other $n - 2$ vertices in its boundary must be saddle points.

6.3 Operations Required to Generate a Minimal Slope Complex

We finish this section with the study of generating a *minimal* slope complex $K[S]$ which is a decomposition of S in the smallest number of slope regions. Observe that this complex satisfies at least that its vertices are only the critical points of S.

First, after contraction of saddle points in each saddle tree, we only have slope regions formed by two edges connecting a saddle point and another edge connecting two vertices in V_+. Figure 4b shows the graph where all saddle points are contracted into a single saddle point.

Lemma 5. *The number of slope regions remains the same if all edges between saddle points are contracted.*

Proof. Edges between saddle points are oriented and, hence, cannot form self-loops, a pre-condition for contraction. The contraction of an edge $(x, y) \in E, x \neq y$, identifies vertices x and y and removes the edge (x, y) from E: The graph after contracting one edge has one less vertex and one less edge. There are no changes in the number of slope regions due to Euler's number is the same after contracting the edge.   □

Finally, the combination of slope regions formed by two triangles may still be simplified. Let us denote by ±-edge an edge in $E_\pm$.

Remark 6. Removing the ±-edge shared by two slope regions produces a new slope region.

The dotted edge in Fig. 5 connects the two extrema. By removing the ±-edge we obtain a new slope region. In this way, the total number of slope regions can be reduced.

Remark 7. The total number of slope regions decreases when two slope regions sharing a ±-edge merge.

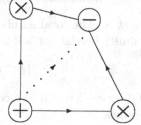

Observe that if the ±-edge bounds the infinite background or it is in the boundary of exactly one slope region, then it cannot be contracted.

Fig. 5. Merging two triangles into one slope region with 4 edges.

Now, we apply the following operations to a given 1-skeleton $G(V, E)$ obtained from a slope complex $K[S]$ with vertices being the critical points of S, to generate a *minimal* slope complex $G'(V, E')$ preserving all the vertices of G and having the minimum number of slope regions satisfying the Euler's formula:

1. Contract all edges in $E_\otimes$.
2. Keep outer ±-edges (i.e., ±-edges in the boundary of the infinite background).

3. Keep all inner ±-bridges (i.e., ±-edges in the boundary of exactly one slope region),
4. Delete all other inner ±-edges (i.e., ±-edges that are not outer ±-edges and are shared by exactly two slope regions).

Lemma 6. *The minimal graph G' is plane, all faces are slope regions and it cannot be further reduced without destroying the property that all faces are slope regions.*

Proof. Deletion of an edge does not change the planarity of a graph. We show that the deletion of a ±-edge merges two slope regions into a new region which is a slope region. A ±-edge of the slope complex G of critical points may be on the boundary or may be a ±-bridge or it may be an inner ±-edge of G. In the first two cases the ±-edge bounds a single slope region. Notice that a bridge need not be an outer edge and is therefore separately mentioned. An inner ±-edge in a slope complex is adjacent to two other minimal slope regions each being a slope region with the same two local extrema. Hence the quadrilateral formed after the removal of the ±-edge is also a slope region. This corresponds to the Quadrangle Lemma in [3]. The argument remains true after the first ±-edge of multiple ±-edges is removed. Therefore all multiple ±-edges can be removed and the merged slope regions still share the same two extrema in a single slope region.   □

Finally we proceed to show the given count of slope regions. We have seen that all edges attached to saddle points do not have any influence on the number of slope regions. All edges between two saddle points can be contracted without reducing the number of slope regions and saddle points cannot form cycles.

Remark 8. The total number of slope regions of G' if the graph doesn't contain any multiple edges or self-loops is:

$$|\{e \in E_\pm \mid e \text{ is not an inner } \pm\text{-edge} \mid \tag{1}$$

Observe that Eq. (1) counts only one slope for outer ±-edges and inner ±-bridges.

7   Conclusions and Future Work

In this paper, we begin with a 2D continuous surface where we define the critical points using local neighborhoods. The surface need not be necessarily piecewise linear or from the smooth category but it should satisfy that critical points are distinct. We then explore the topological properties exhibited by the configuration of given critical points and the space enclosed (slope region) between them. We define slope regions as simply connected components such that any two points in them are connected by a monotonic path. Unlike [3], we allow intersection of monotonic paths and provide a more general topological aspect of the given surface. We show that any 2D continuous surface can be represented as a slope complex and the combinations of different slope regions can be merged to obtain a simplified slope complex. For realization and processing, we use graph-based entities (vertices, edges, and slope regions) similar to [3]. We end the paper by giving the formula to count the minimum number of slope regions required to represent a 2D continuous surface, given the number of critical points.

We started our research on slope regions to better understand the results of our previous work [2] with LBP pyramids where the focus was on critical points. These multiresolution hierarchies of images were built based on a criterion of minimal contrast when merging regions and yields excellent reconstructions with only a very small fraction of image regions. The concept of slope region should enable additional rules to improve the efficiency of our computation.

We know that the partition into slope regions is not unique. What looks as a disadvantage could be used to optimize the receptive field of important critical points and shrink slope regions with "minor importance". Persistent homology should enable further removal of critical points that are due to noise.

Extensions to higher dimensional spaces would establish an interesting connection to frequently used optimization processes. They seek extrema in the space spanned by objective functions. Iterative optimization approaches could be constrained to slope regions leading to the global optimum.

Acknowledgments. This research has been partially supported by MINECO, FEDER/UE under grant MTM2015-67072-P. We thank the anonymous reviewers for their valuable comments.

References

1. Cerman, M., Gonzalez-Diaz, R., Kropatsch, W.: LBP and irregular graph pyramids. In: Azzopardi, G., Petkov, N. (eds.) CAIP 2015. LNCS, vol. 9257, pp. 687–699. Springer, Cham (2015). https://doi.org/10.1007/978-3-319-23117-4_59
2. Cerman, M., Janusch, I., Gonzalez-Diaz, R., Kropatsch, W.G.: Topology-based image segmentation using LBP pyramids. Mach. Vis. Appl. **27**(8), 1161–1174 (2016)
3. Edelsbrunner, H., Harer, J., Zomorodian, A.: Hierarchical Morse - Smale complexes for piecewise linear 2-manifolds. Discrete Comput. Geom. **30**(1), 87–107 (2003)
4. Edelsbrunner, H., Harer, J., Natarajan, V., Pascucci, V.: Morse-Smale complexes for piecewise linear 3-manifolds. Symp. Comput. Geom. **2003**, 361–370 (2009)
5. Edelsbrunner, H., Harer, J.: The persistent Morse complex segmentation of a 3-manifold. In: Magnenat-Thalmann, N. (ed.) 3DPH 2009. LNCS, vol. 5903, pp. 36–50. Springer, Heidelberg (2009). https://doi.org/10.1007/978-3-642-10470-1_4
6. Hatcher, A.: Algebraic Topology. Cambridge University Press, Cambridge (2002)
7. Latecki, L.J., Eckhardt, U., Rosenfeld, A.: Well-composed sets. Comput. Vis. Image Underst. **61**, 70–83 (1995)
8. Kropatsch, W.G., Casablanca, R.M., Batavia, D., Gonzalez-Diaz, R.: On the space between critical points. Submitted to 21st International Conference on Discrete Geometry for Computer Imagery (2019)
9. Peltier, S., Ion, A., Haxhimusa, Y., Kropatsch, W.G., Damiand, G.: Computing homology group generators of images using irregular graph pyramids. In: Escolano, F., Vento, M. (eds.) GbRPR 2007. LNCS, vol. 4538, pp. 283–294. Springer, Heidelberg (2007). https://doi.org/10.1007/978-3-540-72903-7_26

Persistent Homology Computation Using Combinatorial Map Simplification

Guillaume Damiand[1(✉)] and Rocio Gonzalez-Diaz[2]

[1] Univ. Lyon, CNRS, LIRIS, UMR5205, 69622 Villeurbanne Cedex, France
guillaume.damiand@liris.cnrs.fr
[2] Dpto. de Matemática Aplicada I, Universidad de Sevilla, 41012 Seville, Spain
rogodi@us.es
https://liris.cnrs.fr/guillaume.damiand/, http://personal.us.es/rogodi/

Abstract. We propose an algorithm for persistence homology computation of orientable 2-dimensional (2D) manifolds with or without boundary (*meshes*) represented by 2D combinatorial maps. Having as an input a real function h on the vertices of the mesh, we first compute persistent homology of filtrations obtained by adding cells incident to each vertex of the mesh, The cells to add are controlled by both the function h and a parameter δ. The parameter δ is used to control the number of cells added to each level of the filtration. Bigger δ produces less levels in the filtration and consequently more cells in each level. We then simplify each level (cluster) by merging faces of the same cluster. Our experiments demonstrate that our method allows fast computation of persistent homology of big meshes and it is persistent-homology aware in the sense that persistent homology does not change in the simplification process when fixing δ.

Keywords: Persistent homology computation ·
2D combinatorial map · Mesh simplification

1 Introduction

Topological data analysis (TDA) is a relatively new field in computer science. One of the most useful concept in TDA is the one of persistent homology which is an algebraic method for measuring topological features (connected components, voids, cavities, etc) of shapes and functions. Two of the crucial ingredients of persistence are: (1) a *cell complex* to structure the data; and (2) a *filtration* which is a nested sequence of subcomplexes that starts with the empty complex and ends with the whole complex. See [1,2] for initial reports and [3,4] for a modern exposition of the field.

In [5], the authors proposed an efficient algorithm that computes persistent homology for 3D gray-scale images using the Morse-Smale complex previously obtained, which is much smaller than the input data, but with all necessary information. The authors first computed a combinatorial gradient vector field

© Springer Nature Switzerland AG 2019
R. Marfil et al. (Eds.): CTIC 2019, LNCS 11382, pp. 26–39, 2019.
https://doi.org/10.1007/978-3-030-10828-1_3

(GVF) by a process presented in [6]. To do this, the cell complex is decomposed into the lower star of its vertices. The authors then computed persistent homology from the boundary map of the chain complex associated to the Morse-Smale complex induced by GVF.

In [7], we proposed an efficient algorithm for computing the homology of meshes (orientable manifolds with or without boundary), represented by 2D combinatorial maps (which are models of representation of meshes composed by vertices, edges linking two vertices, and 2D faces bounded by a closed path of edges), avoiding the time-consuming step of constructing and modifying boundaries and coboundaries of cells. The process consists of merging faces if they share a common edge, guaranteeing that the structure of combinatorial map and the homology information of the mesh is preserved until the end of the process.

In this paper we extend our work to compute persistent homology of meshes. First, as in [7], a simplification process is made to improve computation time. Now, faces as dispatched in clusters depending on a parameter δ and only faces of the same cluster are merged. For constructing the cluster the following rule is used: two faces are in a same cluster if there is a path of vertices of these two faces of length smaller than δ. At the end of the process, a smaller than the input 2D combinatorial map is obtained. To obtain persistent homology of the simplified mesh, lower-start filtration induced by a function h on its vertices (in our case, h is the height function) is computed. Varying the parameter δ, the filtration varies and also its persistent homology.

The paper is organized as follows. Section 2 recalls the background of the paper regarding combinatorial maps and persistent homology. Section 3 is the main section of the paper and presents our method to compute persistent homology starting from a particular filtration constructed from the height function and a parameter δ. Several experimental and computational results are presented in Sect. 4. Finally, we summarize the paper with a brief discussion about future work in Sect. 5.

2   Preliminary Notions

In this section we recall the needed background of the paper regarding combinatorial maps and persistent homology.

2.1   2D Combinatorial Maps

A 2D combinatorial map [8,9], called 2-map, is a model of representation of a mesh, which is composed by i-cells: vertices or 0-cells associated with points, edges or 1-cells which link two vertices, and faces or 2-cells which are bounded by a closed path.

Two cells are incident if one cell belongs to the boundary of the other one; while two i-cells c_1 and c_2 are adjacent if it exists one $(i-1)$-cell incident to both c_1 and c_2. An edge e is dangling if it is incident to one vertex v such that no other edge than e is incident to v. An edge is isolated if it has no adjacent edge.

An edge incident to two different faces is called *inner*. Such an edge is necessarily not dangling nor isolated. Lastly, an edge is called *border* if it is incident to only one face and if it touches the boundary of the mesh. See Fig. 1(a).

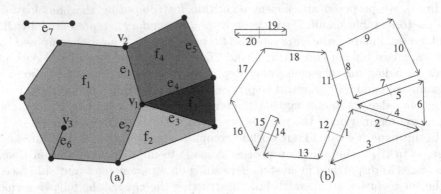

(a)   (b)

Fig. 1. (a) Example of a mesh having 5 faces (the four faces incident to vertex v_1, and the "degenerated one" bounded twice by edge e_7), 14 edges (e_6 is dangling, e_7 is isolated, $\{e_1, e_2, e_3, e_4\}$ are inner and the rest are border) and 12 vertices. (b) The corresponding 2-map has 20 darts. Images taken from [7].

The different elements of a mesh are encoded in a 2-map by *darts* and two mappings between these dart: β_1 and β_2:

β_2: A dart is an orientation of an edge. If an edge separates two faces, it is described by two darts d_1, d_2 in the 2-map linked by β_2 (i.e., $\beta_2(d_1) = d_2$ and $\beta_2(d_2) = d_1$). These two darts represent the two possible orientations of the edge (for example $\beta_2(8) = 11$ and $\beta_2(11) = 8$ in Fig. 1(b)). Each border edge is described by only one dart d in the 2-map, linked by β_2 with a special element $\varnothing$ (cf. for example dart 10 in Fig. 1(b) which describes border edge e_5). β_1: For each dart d, $\beta_1(d)$ is the dart following dart d and belonging to the same face than d (for example $\beta_1(1) = 2$ in Fig. 1(b)). Note that a 2-map is oriented and thus described a given orientation of the mesh.

A dart belongs exactly to one vertex, one edge and one face, and thus each cell of the mesh is described by a set of darts in the 2-map. For example, in Fig. 1(b), vertex v_1 is described by the set of darts $\{2, 5, 8, 12\}$. Note that this is a very important property of 2-map. Even an isolated edge (like e_7 in Fig. 1(a)) belongs to one face (which explain why we have 5 faces and not 4 in Fig. 1(a)).

The different type of edges can be detected in a 2-map thanks to particular configurations of darts and β links (for example an edge is isolated if $\beta_1(\beta_1(d)) = d$, d being of the dart of the edge).

The algorithm presented in this paper for computing persistent homology on meshes used a modified version of Algorithm 1 detailed below which was presented in [7] to compute the minimal 2-map (i.e. with minimal number of cells) describing a given mesh. The algorithm uses two operations on 2-maps: *edge removal* and *edge contraction*. It simplifies a given combinatorial map in its minimal form while preserving all the homology information. The proof that Algorithm 1 preserves homology information is given in [10].

Algorithm 1. Simplification of a mesh (modified version of Algorithm 1 of [7]).

Input: A 2-map M representing the mesh.
Output: The simplified 2-map corresponding to M.

foreach *edge e of* M **do**
    if *e is an inner edge* **then** remove e;

foreach *edge e of* M **do**
    while *e is dangling* **do**
        $e' \leftarrow$ one edge adjacent to e;
        remove e; $e \leftarrow e'$;

foreach *edge e of* M **do**
    if *e is not a loop* **then**
        contract e;

2.2 Persistent Homology

In this subsection we give elementary notions from topology needed to understand the rest of the paper. In particular, we introduce the notion of homology and persistent homology. Precise definitions of homology can be found for example in [11], and definition of persistent homology for example in [4].

Homology can be thought as a method for defining k-dimensional holes (connected components, tunnels, voids) in a given mesh. For example, a 1-cycle is a closed path and a 1-boundary is the boundary of a 2D manifold. Then, 1-homology classes (which represent tunnels) are equivalence classes of 1-cycles modulo 1-boundaries. This concept can be generalized to k-homology classes. Finally, k-homology groups are the groups of k-homology classes.

Persistent homology captures the topological changes occurring in a growing sequence of meshes, called *filtration*. During the growth of a mesh, homology classes of different dimension may appear (be born) and disappear (die). Filtrations are frequently constructed using a real-valued function h on the vertices of the mesh M. For example, the *lower-start filtration* is computed as follows:

– First, order the vertices of M in a non-decreasing way,

$$h(v_1) \leq h(v_2) \leq \cdots \leq h(v_n).$$

– Second, compute the lower-star of a vertex v in M, which is the set of cells of M incident to v whose vertices all have function values at most $h(v)$.

– Define M_i as the union of the lower-star of all vertices of M whose function value is at most $h(v_i)$.
This way, if $h(v_{i-1}) < h(v_i)$ then $M_i \setminus M_{i-1}$ is the set of cells of M having a vertex with function value exactly $h(v_i)$.
And if $h(v_{i-1}) = h(v_i)$ then $M_{i-1} = M_i$.

The lower-star filtration of the mesh M induced by the function h is the sequence of nested meshes:

$$\emptyset = M_0 \subseteq M_1 \subseteq M_2 \subseteq \cdots \subseteq M_{n-1} \subseteq M_n = M.$$

Intuitively, imagine we sweep the mesh M in increasing values of the function h. At any real-value α, we consider the set of cells whose function value on their vertices is below or equal to α. As α increases, this gives us a sequence of subsets of M, growing larger and larger.

The topological evolution along the filtration is expressed by the corresponding sequence of homology groups. When adding the cells in order according to the filtration, new homology classes may born and some of them may later die when they become trivial or merge with another class. If a homology class γ is born at M_i and dies entering M_j then $h(v_j) - h(v_i)$ is the *persistence* of γ. If γ is born at M_i but never dies then its persistence is set to infinity. Homology classes with low persistence are considered noise and the ones that persist are considered features of the mesh.

The information obtained when computing persistent homology can be visualized as a *persistence barcode* which consists of the set of (birth, death) intervals, each interval recording a persistent homology event. The bottleneck distance is used to compare two persistence barcodes corresponding to two different filtrations of the same mesh. Given a bijection η between two persistence barcodes, we take the supremum L_∞-distance[1] between matched points and define the bottleneck distance by taking the infimum over all supremums.

In order to compute persistent homology, in this paper we have implemented a simplified version of the incremental algorithm for computing AT-models given in [12]. Given an ordering of the cells of the mesh, Algorithm 2 computes a triplet (M, H, f) where:

– M is the given mesh (decomposed in cells obtained from the combinatorial map). If σ is a k-cell, then $\partial(\sigma)$ is the set of $(k-1)$-cells in its boundary.
– H is a subset of cells of M called *surviving cells*. Fixed k, the set of all the surviving k-cells together with the addition operation $+$ (here $+$ means the disjoint union of sets) form the group $C_k(H)$ which is isomorphic to the k-dimensional homology group Hk of M.
– $f : C(M) \to C(H)$ maps each k-cell in M to a sum of surviving cells, satisfying that if $a, b \in C_k(M)$ are two homologous k-cycles then $f_k(a) = f_k(b)$. Let M_{σ_i} be the set of cells $\{\sigma_1, \ldots, \sigma_i\}$. Then, in the ith step of Algorithm 2,

---

[1] The L_∞-distance between points $u = (u_1, u_2)$ and $v = (v_1, v_2)$ in the extended plane is $\max\{|u_1 - v_1|, |u_2 - v_2|\}$.

Algorithm 2. Computing persistent homology (Algorithm 2 of [12]).

Input: An ordering of the cells of M: $\{\sigma_1, \ldots, \sigma_m\}$.
Output: Persistent homology.

Initialize $H := \emptyset$ and $f(\sigma_i) := 0$, for $1 \leq i \leq m$.
for $i = 1$ **to** m **do**
   if $f\partial(\sigma_i) = 0$ **then**
       $f(\sigma_i) := \sigma_i$, $H := H \cup \{\sigma_i\}$ (a new homology class was born).
   if $f\partial(\sigma_i) \neq 0$ **then**
       Let $\sigma_j \in f\partial(\sigma_i)$ s t. $j = \max\{$ index$(\mu) : \mu \in f\partial(\sigma_i) \}$
       $H := H \setminus \{\sigma_j\}$ (an old homology class died).
       foreach $x \in M$ *such that* $\sigma_j \in f(x)$ **do**
          $f(x) := f(x) + f\partial(\sigma_i)$.

σ_i belongs to a k-cycle c in $C(M_{\sigma_i})$ if and only if $f\partial(\sigma_i) = 0$. This is why if $f\partial(\sigma_i) = 0$ then a new homology class was born (the one represented by the k-cycle c) and σ_i enters H. Otherwise, if $f\partial(\sigma_i) \neq 0$, then a homology class died, which is equivalent to say that an element of $f\partial(\sigma) \subseteq H$ is removed from H. The element to be removed from H will be the youngest one: $\max\{$ index$(\mu) : \mu \in f\partial(\sigma_i) \}$, being index$(\mu)$ the position of the cell μ in the given ordered list of cells $\{\sigma_1, \ldots, \sigma_m\}$.

In [13] the authors establish a correspondence between the incremental algorithm for computing AT-models given in [12] and the one for computing persistent homology [4]. Since we are only interested in computing the persistence events, we only compute the set H and the map f. See Algorithm 2.

3   Computing Persistence

Our starting point is a subdivision of a mesh M (with or without boundaries) into vertices, edges and faces, and a real-valued function h on the vertices of the mesh.

Our method is based on three steps:

1. Simplification of the 2-map according to a parameter δ;
2. Computation of the lower-star filtration of the simplified mesh;
3. Computation of persistent homology of the given filtration.

Our goal in step 1 is to simplify the 2-map decreasing the number of faces in each level of the filtration in order to improve the computation time in Step 3 which is the more time-consuming step. Observe that persistent homology varies when δ varies since the filtration computed is different. Nevertheless, we have observe in the experiments that our simplification can be seen as a filtering of small persistent homology events.

3.1   2-Map Simplification

In this step, the 2-map is simplified by dispatching the faces into clusters and applying Algorithm 1 with constraints.

First, faces are dispatched into clusters according to the parameter δ. To compute such clusters, vertices of the mesh are ordered in a non-decreasing way by their height values $h(v)$. We assign a height value to each face with is the maximum value of the height of its vertices.

Then in the first cluster we add the first face f in the ordering and all the faces "at distance" less than δ. which means that there exists a path of vertices of these two faces of length smaller than δ. For example, if $\delta = 0$, only one face per cluster is added. If $\delta = 1$ all the faces sharing an edge with f are added. For any $\delta > 1$ all the faces at distance less than or equal to δ to f are added to the cluster. We repeat the process with the next face provided by the ordering that was not included in any cluster. We repeat the process until all faces are in a cluster.

After dispatching the faces in clusters, we apply Algorithm 1 with the following constraints:

- Faces merge (i.e, the inner shared edge e is removed) only if they belong to the same cluster.
- Besides, contrary to Algorithm 1, critical edges (separating faces belonging to two different clusters) are not removed here. Merging faces belonging to two different clusters could lead to loose a persistent event, and this is why we do not merge such faces.
- We do not use the contraction step (last **foreach** in Algorithm 1). Indeed, the simplified 2-map obtained here has several faces, contrary to Algorithm 1 computed without constraints that always produces one face per connected component. For this reason, the number of possible edges to contract is here smaller and thus we have observed no gain (and even sometimes a loss) when using the contraction step comparing to not use it.

3.2   Filtration

The second step in our algorithm for computing persistent homology is to compute the lower-star filtration (see Sect. 2.2) of the simplified mesh SM.

Observed that increasing the value of δ in Step 1 will decrease the different number of SM_j sets (i.e., the number of levels in the filtration), which increases the average number of cells belonging to a same SM_j, as illustrated in Fig. 3 for the Neptune mesh and three different δ values. Note that bigger δ increases the number of simplifications done and thus decreases the size of the simplified combinatorial map. In this case, the persistent homology computed is not the same than the one obtained by the lower-star filtration on the original mesh (they only coincides when $\delta = 0$). Nevertheless, we have seen in our tests that the effect of the parameter $\delta > 0$ is to remove small persistent homology events. However this new possibility gives to users a way to choose a level to analyze a given mesh, while allowing to speed-up the method.

3.3   Computation of Persistent Homology

The last step of our method is the computation of persistent homology of the simplified mesh SM.

We order the cells in SM according to the given filtration and obtain the ordered set of cells $\{\sigma_1, \ldots, \sigma_m\}$ such that if $i < j$ then there exist i', j' such that $i' \leq j'$, $\sigma_i \in SM_{i'}$, $\sigma_j \in SM_{j'}$ and σ_j is not in the boundary of σ_i. We then apply Algorithm 2 to compute persistent homology.

The persistence barcode is stored in a list L with the (birth, death) events as follows: if $\sigma \in M_\ell \setminus M_{\ell-1}$ is born and dies entering $\mu \in M_m \setminus M_{m-1}$, then store (birth, death) in L being birth $= h(v_{i_\ell})$ and death $= h(v_{i_m})$.

Finally, bottleneck distance between different filtrations of the same mesh obtained from different values of δ can be computed to measure the effect of the parameter δ in the persistent homology information obtained.

4   Experiments

We have implemented our algorithm for persistent homology computation by using the CGAL implementation of combinatorial maps [14] and the additional layer, called linear cell complex, which additionally represents the geometry [15]. All our experiments were run on an Intel®i7-4790 CPU, 4 cores @ 3.60 GHz with 32 Go RAM. All the computation times given here are averages of 10 consecutive runs of the same method.

In our tests, we used the six meshes shown in Fig. 2, having between $703, 512$ and $10,000,000$ faces. All these meshes have only one connected component, except *Blade* which has 295 connected components because it contains many small isolated closed meshes inside the blade.

In our experiment, we compared the persistent homology computation of the six meshes for the following values of δ: 0, 1, 2, 4, 8, 16, 32 and 64. For $\delta = 0$, the persistent homology computed is the one of the lower-star filtration induced by the height function on the vertices of the original mesh. When δ increases, the number of faces in a same cluster increases also and thus the combinatorial map becomes more and more simplified. Nevertheless, persistent homology varies since the filtration varies, although differences are "small".

We can see an illustration of the effect of the δ parameter on the size of the different clusters in Fig. 3. The number of cells of the different simplified 2-maps for each value of δ is given in Fig. 4 (average values for the six meshes).

The effect of δ on the computation time is analyzed in Fig. 5 where the six meshes shown in Fig. 2 are used, and our method of persistent homology computation based on the 2-map simplification is ran by using different values of δ. Obviously, computation time decreases while δ increases, since more faces belong to the same cluster, and thus the combinatorial map becomes more and more simplified. We can see that the computation time decreases a lot even for small value of δ which is very interesting. For example, for $\delta = 2$, computation time is divided by 2.75 in average.

(a)

(b)

(c)   (d)   (e)   (f)

	#0-cells	#1-cells	#2-cells	#H0	#H1	#H2
(a) Blade	882,954	2,648,082	1,765,388	295	330	295
(b) DrumDancer	1,335,436	4,006,302	2,670,868	1	0	1
(c) Neptune	2,003,932	6,011,808	4,007,872	1	6	1
(d) HappyBuddha	543,652	1,631,574	1,087,716	1	208	1
(e) Iphigenia	351,750	1,055,268	703,512	1	8	1
(f) ThaiStatue	4,999,996	15,000,000	10,000,000	1	6	1

Fig. 2. The six meshes used in our experiments. The table gives the number of i-cells, #i-cells, and the number of Hi generators, #Hi, for $i = 0, 1, 2$.

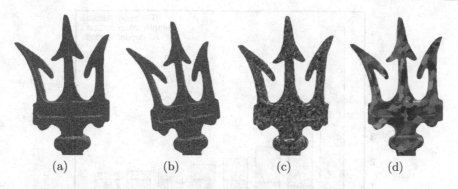

Fig. 3. Effect of the δ parameter on the size of the different clusters for the Neptune mesh, zoom in on the trident. (a) $\delta = 0$. (b) $\delta = 1$. (c) $\delta = 4$. (c) $\delta = 32$.

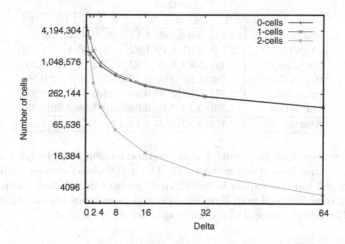

Fig. 4. Number of vertices, edges and faces of the simplified combinatorial maps (in $\log_2$ scale) depending on the value of δ. $\delta = 0$ is the original (non-simplified) 2-map. These numbers are average values for the six meshes.

Table 1. Bottleneck distance between 0-dimensional persistent homology computed on: (1) the lower-star filtration for $\delta = 0$, and (2) the lower-star filtration for different values of δ.

δ	1	2	4	8	16	32	64
Blade	0.64	1	1.53	2.5	3.43	16.25	10.30
DrumDancer	0.10	0.87	0.62	1.25	1.18	3.31	3.31
Neptune	1.10	1.25	1.67	3.08	5.41	8.00	13.41
HappyBuddha	0.00025	0.0005	0.0014	0.0017	0.0024	0.0060	0.010
Iphigenia	0.88	1.19	1.64	2.71	4.51	9.87	19.22
Statuette	0.90	12.37	12.37	12.24	18.39	26.20	27.41

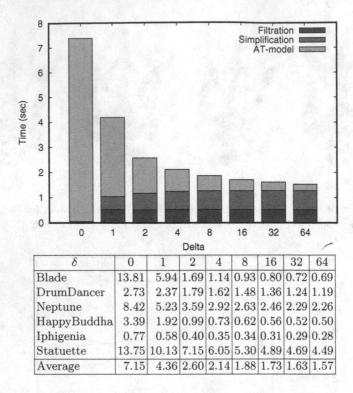

δ	0	1	2	4	8	16	32	64
Blade	13.81	5.94	1.69	1.14	0.93	0.80	0.72	0.69
DrumDancer	2.73	2.37	1.79	1.62	1.48	1.36	1.24	1.19
Neptune	8.42	5.23	3.59	2.92	2.63	2.46	2.29	2.26
HappyBuddha	3.39	1.92	0.99	0.73	0.62	0.56	0.52	0.50
Iphigenia	0.77	0.58	0.40	0.35	0.34	0.31	0.29	0.28
Statuette	13.75	10.13	7.15	6.05	5.30	4.89	4.69	4.49
Average	7.15	4.36	2.60	2.14	1.88	1.73	1.63	1.57

Fig. 5. Computation time (in seconds) of our method by using the patch filtration with increasing δ starting from 0 and going to 64. The graph shows average values for the six meshes, and details time spent in the different parts of the method: computation of the filtration, combinatorial map simplification and persistence computation by using AT-model. The array gives global computation time for each mesh.

We can see in Fig. 6 the effect of δ on the results of the persistent homology computation. First, it should be notice that infinite events are always the same whatever the value of δ is. This is a direct consequence of the fact that the homology of the mesh is preserved by our simplification algorithm. For finite events, we can see that their numbers decrease when δ increase. Indeed, the combinatorial map becomes more and more simplified, and thus the number of cells becomes smaller and smaller (as seen in Fig. 4).

In Table 1 we can see the bottleneck distance with respect to the 0-dimensional persistent homology between the persistence barcodes corresponding to the lower-star filtration and the filtration obtained when varying δ. Table 2 shows the same information for the 1-dimensional persistent homology. To compute the bottleneck distance we used the package TDA of R^2. We can observe that, in general, the distance increases when δ increases and the distance is bounded by the value of δ. Sometimes, δ increases and the distance is a bit lower.

---

[2] https://cran.r-project.org/web/packages/TDA/vignettes/article.pdf.

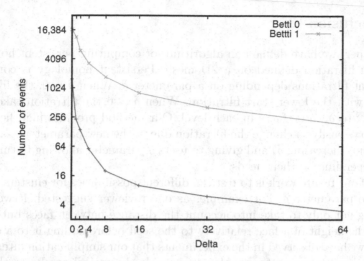

Fig. 6. Number of finite persistence events (in $\log_2$ scale) depending on the value of δ. $\delta = 0$ is the original (non-simplified) 2-map. Betti i is the number of i-homology classes that were born and later died when computing persistent homology, for $i = 0, 1$. These numbers are average values for the six meshes.

This could occurs due to small pockets in the considered mesh. Moreover we can see that in some meshes the effect of δ is more important than in others. See for example Table 1: for Statuette, the difference between the bottleneck distance for $\delta = 0$ and $\delta = 4$ and for $\delta = 0$ and $\delta = 8$ is only $12.37 - 12.24 = 0.13$ which means that we obtain similar persistent homology information when computing persistent homology using $\delta = 8$ instead of $\delta = 4$. Nevertheless, bottleneck distance for $\delta = 0$ and $\delta = 8$ and for $\delta = 0$ and $\delta = 16$ is $18.39 - 12.24 = 6.14$ which means that we could loss important details if we simplify the mesh using $\delta = 16$ instead that $\delta = 8$.

Table 2. Bottleneck distance between 1-dimensional persistent homology computed on: (1) the lower-star filtration for $\delta = 0$, and (2) the lower-star filtration for different values of δ.

δ	1	2	4	8	16	32	64
Blade	0.97	1.0	2.0	4.0	7.0	14.0	21.0
DrumDancer	0.14	0.19	0.38	0.58	1.06	2.40	1.89
Neptune	0.53	1.32	1.82	3.15	5.15	8.28	12.41
HappyBuddha	0.00039	0.00067	0.0010	0.0015	0.0028	0.0034	0.005
Iphigenia	0.6	1.2	1.45	2.87	3.59	9.97	7.27

5   Conclusion

In this paper, we have defined an algorithm for computing persistent homology of a given filtration defined on a 2D mesh. Persistent homology is computed on different filtrations depending on a parameter δ. When $\delta = 0$, the filtration coincides with the lower start filtration. When $\delta > 0$ the filtration takes, proportionally to δ, more faces in each level. Our method provides high flexibility which allows easily to change the filtration due to the new parameter δ, allowing to speed-up (increasing δ) and giving to users a parameter allowing to tune their results depending on their needs.

One of our future work is to test the different possibilities for clusters regarding to the parameter δ. For example, as one reviewer suggested, it would be interesting not only to take into account the distance between faces but also to consider the height of a face relatively to the seed before adding it to a cluster.

Since we have observed in the experiments that our simplification filters small persistent homology events, we plan to provide theoretical results to this new approach stating that the filtration is stable with respect to δ. That is, the bottleneck distance between two filtrations of the same mesh is bounded by a function on δ. We think we can prove it using the classical result of Edelsbrunner et al. on stability of persistence diagrams [4].

Finally, we plan to extend our work to non-orientable manifolds by using the generalized maps package (the non-orientable extension of combinatorial maps) of CGAL. We also would like to define a parallel version of our method: the combinatorial map simplification was already defined in parallel in [7] but we need now to study if it is possible to parallelize some parts of the AT-model computation algorithm. Extension in nD could be given based on the theoretical results for removal and contraction operations in any dimension given in [16,17]. Indeed, all basic tools used in this work, combinatorial maps, removal/contraction operations and AT-model computation, are defined in any dimension.

Acknowledgments. This research has been partially supported by MINECO, FEDER/UE under grant MTM2015-67072-P. We thank the anonymous reviewers for their valuable comments.

References

1. Dey, T.K., Edelsbrunner, H., Guha, S.: Computational topology. In: Advances in Discrete and Computational Geometry. American Mathematical Society, pp. 109–143 (1999)
2. Bern, M.W., et al.: Emerging challenges in computational topology, CoRR cs.CG/9909001
3. Carlsson, G.: Topology and data. Bull. Am. Math. Soc. **46**(2), 255–308 (1999)
4. Edelsbrunner, H., Harer, J.: Computational Topology - An Introduction. American Mathematical Society (2010)
5. Günther, D., Reininghaus, J., Wagner, H., Hotz, I.: Efficient computation of 3D Morse-Smale complexes and persistent homology using discrete Morse theory. Vis. Comput. **28**(10), 959–969 (2012)

6. Robins, V., Wood, P., Sheppard, A.: Theory and algorithms for constructing discrete morse complexes from grayscale digital images. IEEE Trans. Pattern Anal. Mach. Intell. **33**(8), 1646–1658 (2011)
7. Damiand, G., Gonzalez-Diaz, R.: Parallel homology computation of meshes. In: Bac, A., Mari, J.-L. (eds.) CTIC 2016. LNCS, vol. 9667, pp. 53–64. Springer, Cham (2016). https://doi.org/10.1007/978-3-319-39441-1_6
8. Lienhardt, P.: N-dimensional generalized combinatorial maps and cellular quasi-manifolds. Int. J. Comput. Geom. Appl. **4**(3), 275–324 (1994)
9. Damiand, G., Lienhardt, P.: Combinatorial Maps: Efficient Data Structures for Computer Graphics and Image Processing. A. K Peters/CRC Press (2014)
10. Damiand, G., Peltier, S., Fuchs, L.: Computing homology for surfaces with generalized maps: application to 3D images. In: Bebis, G., et al. (eds.) ISVC 2006. LNCS, vol. 4292, pp. 235–244. Springer, Heidelberg (2006). https://doi.org/10.1007/11919629_25
11. Hatcher, A.: Algebraic Topology. Cambridge University Press, Cambridge (2002)
12. Gonzalez-Diaz, R., Real, P.: On the cohomology of 3D digital images. Discrete Appl. Math. **147**(2–3), 245–263 (2005)
13. Gonzalez-Diaz, R., Ion, A., Jimenez, M.J., Poyatos, R.: Incremental-decremental algorithm for computing AT-models and persistent homology. In: Real, P., Diaz-Pernil, D., Molina-Abril, H., Berciano, A., Kropatsch, W. (eds.) CAIP 2011. LNCS, vol. 6854, pp. 286–293. Springer, Heidelberg (2011). https://doi.org/10.1007/978-3-642-23672-3_35
14. Damiand, G.: Combinatorial maps. In: CGAL User and Reference Manual, 3.9 edn (2011). http://www.cgal.org/Pkg/CombinatorialMaps
15. Damiand, G.: Linear cell complex. In: CGAL User and Reference Manual, 4.0 edn (2012). http://www.cgal.org/Pkg/LinearCellComplex
16. Damiand, G., Gonzalez-Diaz, R., Peltier, S.: Removal operations in nD generalized maps for efficient homology computation. In: Ferri, M., Frosini, P., Landi, C., Cerri, A., Di Fabio, B. (eds.) CTIC 2012. LNCS, vol. 7309, pp. 20–29. Springer, Heidelberg (2012). https://doi.org/10.1007/978-3-642-30238-1_3
17. Damiand, G., Gonzalez-Diaz, R., Peltier, S.: Removal and contraction operations in nD generalized maps for efficient homology computation, CoRR abs/1403.3683

Inferring Underlying Manifold of Data by the Use of Persistent Homology Analysis

Rentaro Futagami$^{(\boxtimes)}$ (iD), Noritaka Yamada$^{(\boxtimes)}$ (iD), and Takeshi Shibuya$^{(\boxtimes)}$ (iD)

University of Tsukuba, Tsukuba, Ibaraki, Japan
{futagami,okayasu}@fz.iit.tsukuba.ac.jp, shibuya@iit.tsukuba.ac.jp

Abstract. Inferring underlying manifold of data is one of the important issues for point cloud data analysis. This is accomplished by inferring the topological shape of the underlying manifold. This is done by estimating the number of holes in the underlying manifold in each dimension.

Persistent homology is one of the means of estimating the number of holes in the underlying manifold. Calculating the persistent homology of data determines the size, number, and dimensions of holes produced from data points. However, the number of holes represented through persistent homology is far greater than that in underlying manifold. This problem is caused by noises in a result of calculating persistent homology. Therefore, reducing noises that result from calculating persistent homology is necessary to estimate the number of holes in the underlying manifold.

Conventional methods cannot reduce noises adequately when data are of low density and thus cannot estimate the number of holes in the underlying manifold without manual analysis by experts.

In this study, we propose a new method to estimate automatically the number of holes in the underlying manifolds. We also compare the proposed and conventional methods and show the effectiveness of the former.

Keywords: Persistent homology · Topological data analysis
Underlying manifold · Topological features · Persistent landscape

1 Introduction

Inferring underlying manifold of data is efficient for point cloud data analysis. For example, ensuring that the topology of a graph for an self-organizing map (SOM) is the same as that of the underlying manifold of data is critical [1]. This enable the SOM to preserve the topological relationship among data points. Inferring underlying manifold of data is accomplished by estimating the number and dimensions of "holes" in the underlying manifold and inferring the topological shape based on the same number and dimensions of "holes". A "hole" is defined as a topological feature such as the ring of a donut and a void in a sphere.

© Springer Nature Switzerland AG 2019
R. Marfil et al. (Eds.): CTIC 2019, LNCS 11382, pp. 40–53, 2019.
https://doi.org/10.1007/978-3-030-10828-1_4

The size, number, and dimensions of holes composed of point cloud data are determined by calculating the persistent homology [2,3]. However, in many cases, the number of holes represented through persistent homology is greater than the number of holes in the underlying manifold. This problem is derived from holes in calculation result of persistent homology that are noises composed of cracks among points on a surface of the underlying manifold. Therefore, reducing noises in the calculation result of persistent homology is necessary. Conventional methods [4,5] cannot entirely reduce noises when the density of data points is not sufficiently high to reconstitute the underlying manifold completely. Furthermore, experts must derive the number of holes contained in the underlying manifold from visualized calculation results.

Techniques that accurately and automatically estimate the number of holes in an underlying manifold regardless of the density of data points are required. These are useful for applications of persistent homology.

This study proposes a method to infer holes in an underlying manifold automatically from the calculation results of persistent homology. In a comparison, we show that the proposed method infers holes in an underlying manifold from low-density data more accurately than when using conventional methods.

2   Persistent Homology

Persistent homology is a recently developed novel data analytical tool. It is a method that detects holes that are areas where are surrounded by data points and data points do not exist in point cloud data in a n-dimensional space. Persistent homology is used for not only SOMs but also structural analysis of proteins [6], optimization of sensor networks [7] and natural language processing [8].

Let X be the point cloud data to be calculated using persistent homology, where X is sampled from a D-dimensional manifold $\mathbb{M} \subset \mathbb{R}^D$. In addition, let $h_n(X) = \{(b,d)|b,d \in \mathbb{R}\}$ denote n-dimensional holes in data X derived from calculating persistent homology, where b, d represent birth time and death time of a hole, respectively. Suppose that $(n+1)$-dimensional balls with radius r centering each data point are drawn and r increase, an n-dimensional hole births among balls when $r = b$ and the hole death when $r = d$. A persistent diagram is a graph that maps $h_n(X)$ as coordinate and thus represents persistent homology [9].

Figure 2 is a persistent diagram representing the calculation results of persistent homology shown in Fig. 1. The red triangle in Fig. 2 indicate 1-dimensional holes. An 1-dimensional hole is a loop. In the graph, distances between each point and the diagonal $(d - b)$ represents the size of each hole (known as "persistence"). The larger the persistence is, the larger hole is. Similarly, the green crosses in Fig. 2 indicate 2-dimensional holes, which are enclosed solid voids. The black circles are 0-dimensional holes, which are clusters of data X with d as clustering parameter.

Please see references [2,3] for a more detailed description of persistent homology. This paper describe a method to estimate the number of holes in underlying manifold automatically.

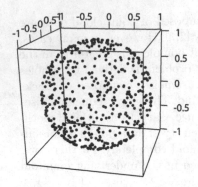

Fig. 1. Example data with sphere topology

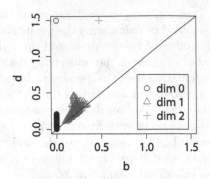

Fig. 2. Persistent diagram of example data with sphere topology (Color figure online)

Table 1. The number of holes in topological shapes

Topological shape	1-dimensional holes	2-dimensional holes
Circle	1	0
Sphere ($\mathbb{S}^2$)	0	1
Torus ($\mathbb{S}^1 \times \mathbb{S}^1$)	2	1

2.1 Inferring Underlying Manifold by Persistent Homology

The topological shape of the underlying manifold of data is inferred by estimating the number of holes in the underlying manifold in each dimension and inferring the topological shape based on the same number of holes. For example, if the underlying manifold have one 1-dimensional hole and two 2-dimensional holes, the topological shape of the underlying manifold is a torus. Table 1 shows a few examples of topological shape and the number of holes in each shape.

In general, the number of holes derived from calculating persistent homology is greater than that in the underlying manifold. Cracks on a surface of underlying manifold produce noises when calculating persistent homology. For example, a sphere that is the underlying manifold of Fig. 1 does not have 1-dimensional hole but only a single 2-dimensional hole. However, in Fig. 2, there are red triangles that indicate 1-dimensional tiny holes on the surface of the sphere. These 1-dimensional holes are noises. The 2-dimensional hole with largest persistence in Fig. 2 corresponds to a hole in the underlying manifold. We call this hole a "cycle" in this paper. We call 1- or more dimensional holes "noise" except holes that are actually in the underlying manifold.

Statistically analyzing the calculation results persistent homology of several samples is effective for data that cannot easily discriminate between cycle and noise [5]. However, obtaining so many data set is not possible in reality. In addition, the computation time and complexity increase depending on the scale of the data. Instead, we used n_{Bi} points subsamples B_i extracted from n_X points data X and calculate persistent homology of these subsamples.

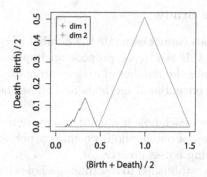

Fig. 3. Example of persistent landscape of data with sphere topology

2.2 Related Works

Several studies have attempted to reduce noises in the calculation results of persistent homology. Fasy et al. proposed a confidence interval of noises in a persistent diagram derived from the Hausdorff distance between subsamples [4]. They also provided proofs of the theorems. However, the calculated confidence interval derived from using this method is a pessimistic threshold based on the premise that extremely many data points exist. Most cycles are considered as noises by this method.

Bubenik focused on that noises tend to have near birth and death time to cycles and slightly smaller persistence than cycles [5]. In addition, they proposed a persistent landscape that maps each coordinate $p = (b, d)$ to a piecewise liner function such that

$$\Lambda_p(t) = \begin{cases} t - b & t \in [b, \frac{b+d}{2}] \\ d - t & t \in (\frac{b+d}{2}, d] \\ 0 & \text{otherwise.} \end{cases} \tag{1}$$

Figure 3 is a graph of Fig. 2 mapped using persistent landscape. Persistent landscape identifies a hole that has the largest persistence in a domain as a cycle and holes that have "lower-ranked" persistence than cycles as noises. We call a local maximum in persistent landscape graphs a "tent". Counting the number of tents indicating "higher-ranked" holes in a persistent landscape graph, as shown in Fig. 3, can derive the number of cycles in Fig. 2. Moreover, excluding "lower-ranked" holes makes analyzing a graph easier. However, determining whether small tents appearing on the sides of large tents in persistent landscape graphs indicate cycles or noises is necessary.

Bubenik also proposed a mean landscape, which average persistent landscape graphs with a confidence interval [5]. The confidence interval is derived from persistent landscape graphs of subsamples of given data. However, mean landscape graphs also have small tents on the sides of large tents, and determining whether small tents indicate noises or cycles is still necessary.

2.3 Purpose of This Study

The conventional methods cannot estimate the number of cycles without manual analysis by experts. In this study, we propose an improved method of persistent landscape to estimate the number of cycles automatically. We compare the proposed method and conventional methods and show the effectiveness of the proposed method.

We determine whether each hole is a noise or a cycle to estimate the number of cycles. In the first step of our method, we employ persistent landscape as an efficient means of reducing noises greatly.

A threshold must be established to discriminate holes based on persistence of holes. This threshold is also necessary to consider holes to be noises in dimensions where there are no cycles or only noises exist. In the second step, we establish a threshold to determine whether or not each dimension has cycles and to discriminate holes. The confidence interval that Fazy et al. proposed is a possible threshold [4]. However, this threshold is not practical. In this study, we propose a new useful threshold.

Finally, as a third step, we analyze small tents that appear on the sides of large tents in persistent landscape graphs. We derive the most probable number of cycles by considering the variation of small tents in persistent landscapes graphs of subsamples statistically. Setting standards to determine whether each small tent indicates a cycle or a noise is difficult. Small tents must be analyzed in somewhat a fuzzy manner. In this study, we propose a statistical method to analyze persistent landscape graphs in a fuzzy manner.

3 Proposed Method

In this section, we describe the method used to estimate the number of cycles automatically. First, we use two kinds of thresholds to reduce noises and then analyze small tents fuzzily by smoothing persistent landscape graphs. The algorithm of the our proposed method to estimate the number of cycles in underlying manifold of data X is shown in Algorithm 1.

With the first threshold, we infer whether data have or do not have a cycle in each dimension. If only extremely small n-dimensional holes exist when compared to the distribution range of data, we determine that the data do not have an n-dimensional cycle. To infer the existence of cycles, we use the mean of the persistence of 0-dimensional holes $h_0(X) = \{(0, d)|d \in \mathbb{R}\}$ as the threshold. The persistence of 0-dimensional holes indicates the proximity of data points. If the variance of data is large, many small holes appear as noises. However, this threshold changes in proportion to the variance of the data and thus accurately identifies the existence of cycles.

Algorithm 1. Estimate the number of n-dimensional cycles in underlying manifold of X

1: **Inputs:**
    X = a input data
    i = the intrinsic dimension of X
    $B_1(X), \cdots, B_m(X)$ = subsamples of X
    $spar \in S$ = a set of smoothing parameters
    n = a target dimension
2: **Outputs:**
    $\hat{b}_n(X)$ = estimated the number of n-dimensional cycles in X

3: **for** $l \leftarrow 1$ to m **do**
4:     $h_k(B_l(X)) \leftarrow$ calculate k-dimensional holes of $B_l(X), (k = 0, \cdots, i)$
5:     $t_1 \leftarrow$ the mean of persistence for all holes in $h_0(B_l(X))$
6:     $t_{\hat{i}} \leftarrow$ the mean of persistence for all holes in $h_k(B_l(X)), (k = 1, \cdots, i)$
7:     $h'_n(B_l(X)) \leftarrow$ find holes by thresholding the persistence of $h_n(B_l(X))$ with t_1
8:     **if** $h'_n(B_l(X)) = \emptyset$ **then**
9:         $\bar{b}_n(B_l(X)) \leftarrow 0$
10:         go to next loop
11:     **end if**
12:     $S(1), S(n) \leftarrow$ calculate formula (2)
13:     $t_2 \leftarrow 2t_{\hat{i}} \times \frac{S(1)}{S(n)}$
14:     **for** $spar \in S$ **do**
15:         $L_n(B_l(X)) \leftarrow$ the n-degree smoothed persistent landscape of $B_l(X)$ with $spar$
16:         $b_{n,spar}(B_l(X)) \leftarrow$ the number of local maxima above t_2 in $L_n(B_l(X))$
17:     **end for**
18:     $\bar{b}_n(B_l(X)) \leftarrow$ the mean of $b_{n,spar}(B_l(X))$ for all $spar \in S$
19: **end for**
20: $\hat{b}_n(X) \leftarrow$ the mean of $\bar{b}_n(B_l(X))$ for all $l(= 1, \cdots, m)$
21: $\hat{b}_n(X) \leftarrow$ round $\hat{b}_n(X)$

At second threshold process, we discriminate between cycles and noises. Let us assume that the number of noises exceeds the number of cycles. The mean of the persistence of all holes can be considered as an approximation of the mean of the persistence of noises. In addition, we assume that the persistence of noises follows the normal distribution in which the mean is equal to that of the persistence of noises. Then, as a threshold, we use the value of the tail opposite to that of the normal distribution whose persistence is 0. In practical term, we double the mean of the persistence of all holes and use this as the threshold. The process of determining the threshold is conducted in each dimension where a cycle is found based on first threshold.

The dimension of hole is larger, the more hole rarely appear. In these threshold process, we must weigh the persistence while considering the dimensions of holes. For example, the unit circle is a 1-dimensional hole and its surface area (circumference) is 2π. In addition, the unit sphere is a 2-dimensional hole and its surface area is 4π. To form both the unit circle and the unit sphere with data

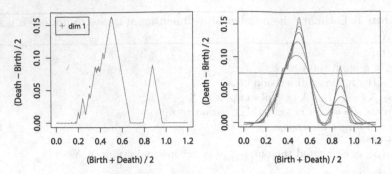

Fig. 4. Example of estimating the number of topological features by proposed method

points having the same variance, the unit sphere must have twice as many data points as the unit circle. The surface area of an n-dimensional hypersphere is given by

$$S(n) = \frac{2\pi^{\frac{n+1}{2}}}{\Gamma(\frac{n+1}{2})}. \tag{2}$$

Therefore, to compensate the differences in easiness of appearing of holes that depend on the dimensions of holes, the n-dimensional persistence must be multiplied by $S(n)/S(1)$.

Third, we reduce noises that appear with cycles in persistent landscape graphs. Let $b_n(X)$ be the number of tents in the n-degree persistent landscape graph, where the number of times that $\mathrm{sgn}(\Lambda_p(t))$ become 0 when t starts from 0 to infinity in Eq. 1 and $b_n(X)$ is the estimated number of n-dimensional cycles in data X. However, $b_n(X)$ is not always correct because it may contain small tents that are on the sides of large tents and that are formed by noises. Analyzing small tents on the sides of large tents is difficult. A universal standard for analyzing small tents has not been proposed. Whether small tents indicate cycles or noises is typically determined in a subjective manner.

To discriminate between cycles and noises automatically, we smooth persistent landscape graphs using various smoothing parameters. We consider that tents that exist even after high smoothing indicate cycles. We fit a cubic smoothing spline using B-spline based on a generalized additive model (GAM) to perform smoothing [10]. Concretely, we find $f(x)$ that minimize σ in Eq. 3:

$$\sigma = \sum_{i=1}^{n} (y_i - f(x_i)) + \lambda \int \{f''(x)\}^2 dx. \tag{3}$$

λ is given by

$$\lambda = \rho * 256^{3*spar-1}, \tag{4}$$

and ρ is given by

$$\rho = \frac{\sum_{i=1}^{n} \{B_i(x_i)\}^2}{\sum_{i=1}^{n} \int \{B_i''(t)\}^2 dt}, \tag{5}$$

where $B_k(\cdot)$ is k-th B-spline. We smooth one persistent landscape graph several times, changing a parameter $spar$ in Eq. 4 sequentially, such as $spar = \{0, 0.1, \ldots, 1\}$. This smoothing process estimates the number of cycles $\bar{b}_n(X) = \sum_{spar \in S} \frac{b_{n,spar}(X)}{||S||}$, where S is a set of $spar$ and $b_{n,spar}(X)$ is b_n of the smoothed persistent landscape graph with the smoothing parameter $spar$.

Figure 4 shows an example of smoothing by fitting cubic smoothing splines. The persistent landscape graph shown on the left side of Fig. 4 has a large tent with small tents to its side. In addition, a medium tent appears a little above threshold 0.0075. The right side of Fig. 4 shows the result of smoothing the persistent landscape graph shown in Fig. 4 with the parameter $spar \in \{0, 0.1, \ldots, 1\}$. In Fig. 4, the small tents on the side of the large tent are merged. The medium tent becomes small and lower than the threshold as smoothing parameter increase. However, the large tent remains. We smooth the persistent landscape graph with each smoothing parameter and count the number of tents above the threshold in each smoothed persistent landscape graph. We then average the number of tents in each smoothed persistent landscape graph and obtain a mean of 2.27. We consider this mean to be the number of cycles in this persistent landscape graph. In Sect. 4, we smooth persistent landscape graphs using same smoothing parameter $spar \in \{0, 0.1, \ldots, 1\}$.

Finally, we describe a statistical method to estimate the number of cycles using subsample. Subsampling yield a smaller size of data than that of the original data and enable faster calculation of persistent homology. In addition, subsampling enables us to use the many calculation results of persistent homology. When subsampling excludes some data points, the persistence of existing holes in the data rarely changes. Therefore, when using subsampling, the number $\hat{b}_n$ of n-dimensional cycles in data X is given as

$$\check{b}_n(X) = \sum_{B(X) \in \mathcal{B}} \frac{\bar{b}_k(B(X))}{||\mathcal{B}||}, \tag{6}$$

$$\hat{b}_n(X) = \lfloor \check{b}_n(X) \rfloor + \lfloor 2(\check{b}_n(X) - \lfloor \check{b}_n(X) \rfloor) \rfloor, \tag{7}$$

where $\mathcal{B}$ is a set of $B(X)$, and $B(X) = \{x_i | i \in I \subset (1, \cdots, n_X), I \overset{i.i.d}{\sim} p(I)\}$ represents subsampled n_B points without overlap from n_X points data X according to the uniform probably distribution $p(I) = \frac{1}{n_X C_{n_B}}$.

We can now summarize the proposed method. First, we subsample the given data to obtain some subsamples. Second, we estimate the number of cycles in each subsample through the two threshold and smoothing processes previously described. Third, we average the estimated number of cycles in the subsamples. We then round off the mean of the number of cycles in the subsamples. Finally, we consider the rounded mean to be the number of cycles in the given data. The topological shape of the underlying manifold is inferred based on the estimated number of cycles by the proposed method.

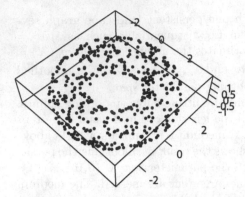

Fig. 5. Experimental data with the torus topology

Table 2. Settings of experiment

Data set	100
Major radius R	2.5
Minor radius r	1
Data points	500
Subsamples	10
Points of a subsample	300
Smoothing parameter	$\{0, 0.1, \ldots, 1\}$

3.1 Restriction of the Proposed Method

Most practical cases demand to estimate the number of only 1- and 2-dimensional cycles. The proposed method is effective for practical cases.

However, like manual analysis by human experts, the proposed method is difficult to infer that the underlying manifold has no cycles when data is too low density. For example, a 400 points spherical data set sampled uniformly from the unit sphere is inferred that it has one 1-dimensional cycle and one 2-dimensional cycle by the proposed method using 240 points subsamples. 1-dimensional cycles do not exist in the unit sphere actually.

We could not examine that the proposed method is effective at how much density of data because of the computational complexity of persistent homology in the large scale data.

4 Experiments

We compared the proposed method with conventional methods that calculate the confidence interval of noises in a persistent diagram [4], persistent landscape, or mean landscape [5].

We compared the proposed and conventional methods based on their ability to accurately estimate the number of cycles in some data. We calculated the persistent homology using optimum parameters to estimate the number of cycles. These optimum parameters are set to increase the computation speed; they are easy to set if users ignore limits of computational resources.

4.1 Torus

Figure 5 shows comparatively low-density data on the torus. The underlying manifold of data shown in Fig. 5 has two 1-dimensional cycles and one 2-dimensional cycle. We estimated the number of cycles in 100 set of data sampled

Table 3. Estimates of the number of 1-dimensional cycles

Cycles	0	1	2	3	4	Success rate
Confidence interval	100	0	0	0	0	0%
Persistent landscape	0	6	81	10	3	81%
Mean landscape	0	11	80	9	0	80%
Proposed method	0	10	90	0	0	90%

Table 4. Estimates of the number of 2-dimensional cycles

Cycles	0	1	2	Success rate
Confidence interval	100	0	0	0%
Persistent landscape	42	58	0	58%
Mean landscape	100	0	0	0%
Proposed method	43	57	0	57%

from the uniform distribution on the torus, as shown in Fig. 5. Table 2 shows the settings for this experiment. Tables 3 and 4 show the comparative results of the estimations using the proposed and conventional methods.

Table 4 shows that both the proposed method and persistent landscape accurately estimated the number of cycles from many set of data. No significant difference was found between the results of the proposed method and persistent landscape. Two explanation can be given for these results. One is that 2-dimensional cycles are difficult to detect in the data because the persistence of 2-dimensional cycles is originally small. Another is the bias in the variance of data. The method employing the confident intervals inferred that all holes are noises and estimated an incorrect number of cycles in all set of data. Mean landscape seemed to estimate an incorrect number of cycles in many set of data because the persistence of 2-dimensional cycles became smaller than the original persistence after the persistence was averaged.

Table 3 shows that the proposed method, persistent landscape, and mean landscape all estimated the accurate number of 1-dimensional cycles in most set of data. Persistent landscape sometimes estimated that the underlying manifold had three or four 1-dimensional cycles after it counted the number of small tents formed by noises. Mean landscape rarely counted the number of small tents formed by noises as compared to persistent landscape. However, mean landscape also estimated that the underlying manifold had three 1-dimensional cycles in some data.

By contrast, the proposed method did not estimate that the underlying manifold had three or more 1-dimensional cycles. In addition, the proposed method could determine whether small tents indicated cycles or noises and it estimate the number of cycles accurately.

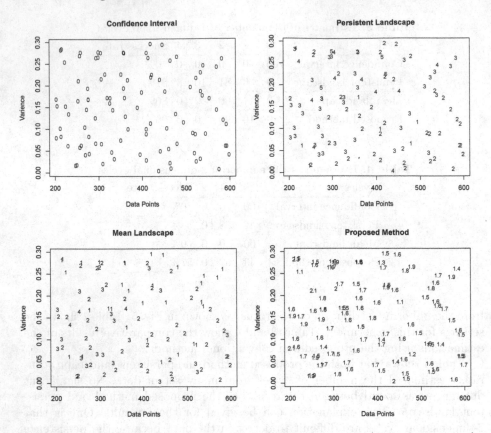

Fig. 6. Estimated number of 1-dimensional cycles in noisy toruses (Color figure online)

4.2 Noisy Torus

Second, we estimated the number of cycles in 100 set of noisy data. Concretely, we sampled from the uniform distribution on the torus. The major radius and the minor radius of the torus are 2.5 and 1, respectively. The number of data points was determined randomly based on the uniform distribution of the interval $[200, 600]$ for every set of data. In addition, we added Gaussian noise to each data point on the torus. The variance of Gaussian noise was determined the uniform distribution of the interval $[0.0, 0.3]$ for every set of data.

Figures 6 and 7 show the results of this experiment. The vertical and horizontal axes of graphs represent the number of data points and the variance of Gaussian noise, respectively. The plotted numbers in the graphs represent the estimated number of cycles in each set of data. The correct and incorrect number of cycles are colored in red and blue, respectively. The results of the proposed method as shown in Fig. 6 (bottom right) and Fig. 7 (bottom right) are represented with real numbers as a result of statistical analysis.

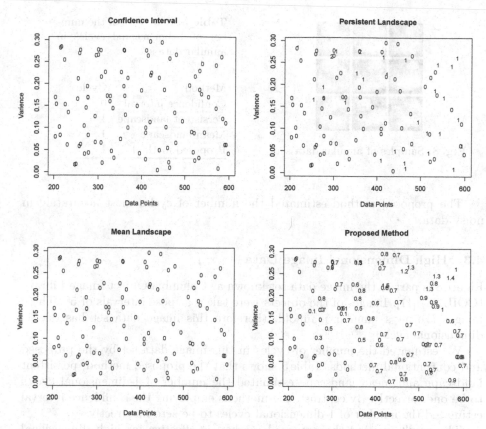

Fig. 7. Estimated number of 2-dimensional cycles in noisy toruses (Color figure online)

Figure 6 shows that the proposed method estimated the correct number of 1-dimensional cycles in most set of data. A success rate of the proposed method was 86% in 1-dimensional cycles. By contrast, both persistent landscape and mean landscape estimated the correct number of cycles in only 42% and 45% the data sets, respectively. The method employing the confidence interval estimated the incorrect number of cycles in all set of data.

Figure 7 shows that the proposed method estimated the incorrect number of 2-dimensional cycles in small sized data. The reason for this result is that 2-dimensional holes cannot be formed when only a few data points exist. However, the proposed method estimated the correct number of 2-dimensional cycles in most set of data that had more than 400 points. A success rate of the proposed method was 61% in 2-dimensional cycles.

By contrast, persistent landscape estimated the incorrect number of cycles in many set of data. A success rate of persistent landscape was 27%. The methods employing the confidence interval or mean landscape estimated the incorrect number of cycles in all set of data.

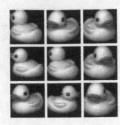

Fig. 8. Samples of annular data set

Table 5. Estimates of the number of 1-dimensional cycles in annular data set

Method	Cycle
Confidence interval	0
Persistent landscape	1
Mean landscape	1
Proposed method	1

The proposed method estimated the number of cycles most accurately in noisy data.

4.3   High Dimensional Image Data

Figure 8 is part of the image data set known as Columbia Object Image Library (COIL-20) [11]. Images of the objects were taken at pose intervals of 5°. These images are on a annular manifold, therefore this image data set has one 1-dimensional cycle.

We estimated the number of cycles in this image data set by the proposed and conventional methods. Table 5 shows that the proposed method, persistent landscape and mean landscape estimated the number of 1-dimensional cycles to be one correctly. By contrast, the methods employing the confidence interval estimated the number of 1-dimensional cycles to be zero incorrectly.

The result insisted the proposed method is effective for high dimensional data.

As the aforementioned results indicate, the proposed method estimates the number of cycles more accurately than conventional methods. Estimation the number of cycles as a real number using subsamples and smoothing enables the proposed method to help inferring the underlying manifold like human experts. Therefore, we get not a single inferring result but several inferred candidates of the underlying manifold with the likelihood.

5   Conclusion

In this study, we proposed the method to estimate the number of cycles in the underlying manifold automatically.

We then estimate the number of cycles in the torus without noise, the noisy torus and the high dimensional data by the proposed method and conventional methods. As a result, the proposed method estimated more accurately the number of cycles than conventional methods regardless noise and extrinsic dimension of data. Using subsamples and smoothing enables the proposed method to estimate the number of cycles with high likelihood. The topological shape of the underlying manifold can be inferred based on the estimated number of cycles.

References

1. Futagami, R., Shibuya, T.: A method deciding topological relationship for self-organizing maps by persistent homology analysis. In: Proceedings of SICE Annual Conference 2016, pp. 1064–1069 (2016)
2. Zomorodian, A., Carlsson, G.: Computing persistent homology. Discret. Comput. Geom. **33**(2), 249–274 (2005)
3. Edelsbrunner, H., Harer, J.: Persistent homology-a survey. Contemp. Math. **453**, 257–282 (2008)
4. Fasy, T.B., Lecci, F., et al.: Confidence sets for persistence diagrams. Annu. Stat. **42**(6), 2301–2339 (2014)
5. Bubenik, P.: Statistical topological data analysis using persistent landscapes. J. Mach. Learn. Res. **16**(1), 77–102 (2015)
6. Gameiro, M., et al.: A topological measurement of protein compressibility. Jpn. J. Ind. Appl. Math. **32**(1), 1–17 (2013)
7. Zhang, W., et al.: An optimized degree strategy for persistent sensor network data distribution. In: Euromicro International Conference on Parallel, Distributed and Network-Based Processing (2012)
8. Zhu, X.: Persistent homology: an introduction and a new text representation for natural language processing. In: Proceedings of the Twenty-Third International Joint Conference on Artificial Intelligence (2013)
9. Steiner, D.C., Edelsbrunner, H., Harer, J.: Stability of persistence diagrams. Discret. Comput. Geom. **37**(1), 103–120 (2007)
10. Hastie, T.J., Tibshirani, R.J.: Generalized Additive Models, 1st edn. Chapman & Hall/CRC, Boca Raton (1990)
11. Nene, S.A., Nayar, S.K., Murase, H.: Columbia object image library (COIL-20). Technical report, No. CUCS-005-96 (1996)

Gaps and Well-Composed Objects
in the Triangular Grid

Lidija Čomić[✉]

Faculty of Technical Sciences, University of Novi Sad, Novi Sad, Serbia
comic@uns.ac.rs

Abstract. We extend the notion of a gap from the square to the triangular grid, and we propose a possible classification of gaps in this grid. We give four definitions of well-composed objects in the triangular grid by translating the existing definitions of such objects in the square grid. We show that these definitions in the triangular grid are equivalent, as they are in the square grid.

We give a formula relating the number of gaps of different types in an object in this grid with the number of boundary cells in the object, as well as three short intuitive proofs of this formula.

Keywords: Digital topology · Triangular grid · Gaps
Well-composedness

1 Introduction

There are three regular grids in the plane, inducing the tiling of the plane into regular triangles, squares or hexagons. Although the square grid remains the most popular in the literature, the two alternative grids have also been widely investigated in different frameworks, such as topology-preserving transformations [15,16,21,26,27,43], computation of the Euler characteristic [6,24,41], analytical [17,20,37] or computational [38,39] geometry, tomography [36], topological/combinatorial coordinate systems [25,35], distance transform [1,2] and neighborhood sequences [18], to name just a few. Each grid has specific geometric and topological properties. One of such topological properties is well-composedness [31,32] of binary objects in the given grid. Closely related is the notion of gaps [9,11], their classification and their number in non-well-composed objects. Despite this close relation, the two notions have been defined and studied independently in the literature, and only in the case of the square (and the nD cubical) grid.

A gap in an object in the square grid is a vertex v incident to exactly two black squares and exactly two white ones, the black and white squares alternating cyclically around v. Intuitively, it is a location in O through which a discrete curve can penetrate [7].

© Springer Nature Switzerland AG 2019
R. Marfil et al. (Eds.): CTIC 2019, LNCS 11382, pp. 54–67, 2019.
https://doi.org/10.1007/978-3-030-10828-1_5

There are several approaches to well-composedness in the square grid. A well-composed object can be defined based on

- the absence of certain (forbidden) configurations of squares in 2×2 blocks of squares, containing two diagonally placed black squares and two diagonally placed white ones,
- the manifoldness of its topological boundary,
- the manifoldness of its combinatorial boundary,
- the independence of the connectedness of the object and its complement on the chosen adjacency relation.

Note that a forbidden configuration occurs exactly when the central vertex of the configuration is a gap. These four definitions of a well-composed object agree in the 2D square grid [5]. Thus, a gap in an object O in this grid is a vertex at which O fails to be well-composed.

Well-composed objects were also defined in the 3D [31] (and nD [3]) cubical grids. The problem of repairing non-well-composed objects in the 3D cubical grid, i.e., of transforming such objects into well-composed ones, has received a lot of attention recently [4,14,22,23,40,42].

We define a gap in an object O in the triangular grid based on the number of boundary edges incident to the vertex, i.e., as a central vertex of certain configurations of black and white triangles. We propose a classification of gaps in the triangular grid based on the 'degree of non-manifoldness' of O at such vertices. We translate the four definitions of well-composed objects from the square to the triangular grid and we show that these definitions are equivalent also in the triangular grid. We give a formula for the number of gaps in O, as an extension of the known formula for the number of gaps in the square grid.

To summarize, the contributions of this paper are:

- an extension of the notion of gaps from the square to the triangular grid, and their tentative classification,
- a translation of the notion of well-composed objects from the square to the triangular grid, with four different definitions of such objects,
- a proof of equivalence of the four definitions,
- three short and intuitive proofs of a formula expressing the number of gaps in an object O through the number of boundary edges and vertices in O.

2 Preliminaries

We give some basic notions on the regular grids in 2D [16,28], the Euler characteristic [29], gaps and well-composed objects in the square grid [5,32] and the number of gaps in the non-well-composed such objects [9,11].

2.1 Regular Grids in the Plane

The only regular grids in the plane are the triangular, square and hexagonal grids, which induce the tessellations of the plane into regular triangles, squares and hexagons, respectively. The triangles, squares and hexagons are called pixels. Each triangle, square and hexagon is bounded by three, four and six edges and vertices, respectively. Each edge (in all three grids) is bounded by two vertices, and bounds two pixels. Each vertex in the triangular, square and hexagonal grid bounds six, four and three pixels and edges, respectively.

Different types of adjacency relation are defined between the pixels in these grids, depending on their intersection. Two pixels are 1-adjacent (edge-adjacent) if they share an entire edge; they are (strictly) 0-adjacent (vertex-adjacent) if they share (only) a vertex.

Each triangle is 1-adjacent to three triangles, one across each of its edges. It is strictly 0-adjacent to another nine triangles, three across each of its vertices. Each square is 1-adjacent to four squares, and is strictly 0-adjacent to other four squares. Each hexagon is 1-adjacent to six hexagons. Two hexagons that share a vertex, share also an entire edge. Thus, there is no strict 0-adjacency in the hexagonal grid.

A (binary) object O in these grids is a finite set of pixels in the grid. The associated cell complex [30] is denoted as Q. It consists of all pixels in O and all their edges and vertices. (Recall that a k-cell is a homeomorphic image of a k-ball. A cell complex is a collection Q of cells that fit nicely together: the boundary of each cell in Q, and each non-empty intersection of two cells in Q is composed of cells in Q.) The number of k-cells in Q is denoted as c_k. Thus, c_2 is the number of pixels, c_1 is the number of edges and c_0 is the number of vertices in Q. The pixels in O are called black (object), those in the complement O^c of O are called white (background).

A boundary edge (vertex) in Q is an edge (vertex) incident both to a pixel in O and a pixel in O^c. An interior edge (vertex) is incident to pixels in O only. The number of interior and boundary k-cells in Q is denoted as c'_k and c^*_k, $k = 0, 1$, respectively, and

$$c_k = c'_k + c^*_k.$$

Two pixels p and q in an object O are 1-connected (0-connected) in O, if there is a sequence of pixels in O, starting at p and ending in q, such that any two consecutive pixels in the sequence are 1-adjacent (0-adjacent). The 1-connected components (0-connected components) of O are maximal subsets of O with respect to the chosen connectedness. The number of 1-connected (0-connected) components is denoted c^1 (c^0). A hole is a finite connected component of the complement O^c of O. If connected components of O are defined with respect to 1-adjacency, then holes are defined with respect to 0-adjacency and vice versa. (In the hexagonal grid, both the connected components and holes are defined with respect to 1-adjacency.) The number of holes with 1-adjacency for O and 0-adjacency for O^c is denoted h^0. Similarly, the number of holes with 0-adjacency for O and 1-adjacency for O^c is denoted h^1.

We say that a cell complex Γ (a finite set of pixels, edges and vertices in a grid that forms a cell complex) is connected if for any two cells in Γ there is a sequence of cells in Γ starting at one and ending at the other cell, such that two consecutive cells share at least one common vertex.

2.2 The Euler Characteristic of 2D Digital Objects

The Euler characteristic is one of the basic topological descriptors of objects (images or shapes). It can be defined in two equivalent ways:

- through the topological properties related to connectedness of O, or
- through the number of cells in the complex Q associated to O.

In both cases, its value depends on the chosen adjacency relation. We denote as $\chi^\alpha(O)$ the Euler characteristic of O with α-adjacency for black pixels and $(1 - \alpha)$-adjacency for white pixels, $\alpha = 0, 1$. The Euler characteristic $\chi^\alpha(O)$ of a 2D digital object O with α-adjacency, $\alpha = 0, 1$, is equal to the number of connected components of O minus the number of holes in O, i.e.,

$$\chi^\alpha(O) = c^\alpha - h^{1-\alpha}.$$

For an object O with 0-adjacency (and for the associated complex Q), the Euler characteristic can be computed as the alternating sum of the number c_i of i-cells in Q, $i = 0, 1, 2$, i.e.,

$$\chi^0(O) = \chi^0(Q) = c_0 - c_1 + c_2.$$

2.3 Gaps and Well-Composed Objects in the Square Grid

A gap in an object O in the square grid occurs at a vertex v incident to two black and two white squares, that alternate cyclically around v. There is only one type of gaps in the square grid, due to the restricted number (four) of squares incident to a vertex in this grid.

There are several equivalent definitions of well-composed objects (sets or images) in the square grid [3,5]. A 2D object O in the square grid is well-composed if:

- the object O contains no forbidden configurations, i.e., (boundary) vertices incident to exactly two strictly vertex-adjacent black squares (and to two strictly vertex-adjacent white ones), that is, O contains no gaps,
- the topological boundary $bd\,O$ of O (the set of points in the plane for which every neighborhood has a nonempty intersection both with the object O and with its complement O^c) is a topological 1-manifold (a topological space in which every point has a neighborhood homeomorphic to the open unit interval, i.e., a union of disjoint curves),
- the combinatorial boundary ∂O of O (the cell complex composed of boundary edges and vertices of O) consists of discrete 1-surfaces [19] (every edge is incident to exactly two distinct vertices and every vertex is incident to exactly two distinct edges),
- the connected components of O and of the complement O^c of O do not depend on the chosen adjacency relation (0- or 1-adjacency), i.e., each 0-connected component of O (and of O^c) is also a 1-connected component.

2.4   The Number of Gaps in the Square Grid

Several formulae have been proposed for the number g of gaps in a binary object O in the square grid [8, 9, 13]. One of them [9] expresses g in terms of the boundary cells in the associated complex Q as

$$g = c_1^* - c_0^*,$$

or equivalently as

$$g = c_1 - c_1' - c_0 + c_0'.$$

In higher dimensions, different types of gaps can be defined. Formulae for the number of 1-gaps in 3D [33] and $(n-2)$-gaps in nD [7, 11, 34] binary objects in the cubical grid have also been proposed.

3   Gaps in the Triangular Grid

Up to rotation and symmetry, there are thirteen different configurations of the six triangles around a vertex v in the triangular grid, as illustrated in Fig. 1. We classify the vertices in O based on the number of boundary edges in O incident to the vertex.

Fig. 1. The possible configurations (up to rotation and symmetry) of black and white triangles around a vertex in the triangular grid: (a) non-boundary vertices (interior or exterior to the object O); (b) non-gap boundary vertices; (c) simple gaps; (d) double gaps.

The vertices incident to six black triangles are *interior* vertices of O, and those incident to six white triangles are *exterior* vertices of O (they are interior vertices of O^c). Interior and exterior vertices are illustrated in Fig. 1(a). Other vertices are boundary vertices of O.

We call the boundary vertices incident to exactly two boundary edges *non-gap* boundary vertices of O. (Such vertices in the square grid are called also totally boundary cells [7,11] or nubs [34].) The six triangles incident to a non-gap boundary vertex can be grouped in two sets of contiguous triangles: a set of 1-connected black triangles and a set of 1-connected white ones. The possible configurations of black and white triangles around a non-gap vertex v are illustrated in Fig. 1(b).

We call the boundary vertices incident to exactly four boundary edges *simple* gaps. The six triangles incident to a simple gap can be grouped in four sets of contiguous triangles: two sets of 1-connected black triangles and two sets of 1-connected white ones. The possible configurations of black and white triangles around a simple gap v are illustrated in Fig. 1(c).

We call the boundary vertices incident to exactly six boundary edges *double* gaps. The three black and three white triangles incident to a double gap v alternate cyclically around v, as illustrated in Fig. 1(d).

Thus, a gap is a vertex incident to at least two strictly 0-adjacent black triangles that are not 1-connected in the set of black triangles incident to v (and to at least two strictly 0-adjacent white triangles that are not 1-connected in the set of white triangles incident to v).

We denote the number of non-gap boundary vertices as $\bar{g}$, the number of simple and double gaps as $g^{(1)}$ and $g^{(2)}$, respectively, and the total number of gaps as g. Thus,

$$g = g^{(1)} + g^{(2)}.$$

Let us consider a cell complex Q', obtained from the complex Q associated with an object O by deleting a gap v, together with all the black triangles and all their edges that are incident to v. Let us consider how the topology of Q' changes when we reintroduce the deleted cells.

At each simple gap v, either

– two connected components of Q' meet (and merge) at the vertex v, or
– one connected component of Q' meets itself forming a hole (a loop),

in both cases decreasing the Euler characteristic $\chi^0(Q')$ by 1, i.e.,

$$\chi^0(O) = \chi^0(Q) = \chi^0(Q') - 1.$$

Each double gap is incident to exactly six boundary edges. The three black and three white triangles incident to a double gap alternate cyclically around the gap.

At each double gap v, either

– three connected components of Q' meet and merge at v, or
– one component forms a loop and merges with another component, or

– one component meets itself two times, forming two independent loops, or
– one component meets itself three times, forming three dependent (and two independent) loops.

In any case, the Euler characteristic $\chi^0(Q')$ is decreased by 2 at a double gap, i.e.,

$$\chi^0(O) = \chi^0(Q) = \chi^0(Q') - 2.$$

Thus, when counting gaps, we will count double gaps twice, and we will give a formula for the number $g + g^{(2)} = g^{(1)} + 2g^{(2)}$.

4   Well-Composed Objects in the Triangular Grid

We give four definitions of well-composed objects in the triangular grid, analogous to the definitions of well-composed objects in the square grid, and we show that the four definitions are equivalent.

Definition 1. *An object O in the triangular grid is well-composed if it has no gaps (i.e., if all its boundary vertices are non-gaps).*

Definition 2. *An object O in the triangular grid is well-composed if its topological boundary is a topological 1-manifold.*

Definition 3. *An object O in the triangular grid is well-composed if its combinatorial boundary ∂O consists of discrete 1-surfaces.*

Definition 4. *An object O in the triangular grid is well-composed if each 0-connected component of O (and of the complement O^c of O) is also a 1-connected component.*

Proposition 1. *Definitions 1 and 2 are equivalent, i.e., an object O in the triangular grid has no gaps if and only if its topological boundary bd O is a topological 1-manifold.*

Proof. ($\Rightarrow$) If the object O has no gaps, then each boundary vertex is a non-gap boundary vertex and is incident to two boundary edges. The topological boundary $bd\ O$ of O is composed of boundary edges and vertices in O, considered as a set of points in the plane. The points in the interior of boundary edges have a neighborhood homeomorphic to the open unit interval, e.g. the open edge itself. For the end points of the edges (the boundary vertices), such neighborhood is e.g. the union of the two incident (open) edges and the vertex itself.

($\Leftarrow$) Let the object O contain at least one gap v. By inspecting the possible configurations of boundary edges around v (see Fig. 1(c) and (d)), we see that the topological boundary $bd\ O$ of O is not a topological 1-manifold at v (as each gap v is incident to four or six boundary edges).

Proposition 2. *Definitions 1 and 3 are equivalent, i.e., an object O in the triangular grid has no gaps if and only if its combinatorial boundary ∂O consists of discrete 1-surfaces.*

Proof. ($\Rightarrow$) The combinatorial boundary ∂O of an object O in the triangular grid is a cell complex composed of boundary edges and vertices in O. If the object O has no gaps, then each boundary vertex in O, being a non-gap vertex, is incident to exactly two boundary edges in ∂O. Each (boundary) edge is a straight line segment, and thus it is incident to exactly two distinct (boundary) vertices. Thus, ∂O is composed of discrete 1-surfaces.

($\Leftarrow$) Conversely, if ∂O is composed of discrete 1-surfaces, then each vertex in ∂O is incident to exactly two (boundary) edges in ∂O. Thus, O has no gaps, since each gap is incident to four or six (and not two) boundary edges.

Proposition 3. *Definitions 1 and 4 are equivalent, i.e, an object O in the triangular grid has no gaps if and only if each 0-connected component of O (and of the complement O^c of O) is also a 1-connected component.*

Proof. ($\Rightarrow$) If there is a 0-connected component in O that is not a 1-connected component (and similarly for the complement O^c of O), then there are two black triangles in this component such that for each sequence of 0-adjacent black triangles connecting them, at least one adjacency is strict. Let v be one such vertex incident to two strictly 0-adjacent black triangles t_1 and t_2 in the same 0-connected component of O. The triangles t_1 and t_2 are not 1-connected in the set of black triangles incident to v. This implies that there are at least two sets B_1 and B_2 of contiguous (1-connected) black triangles around v (incident to v), and at least two such sets of white triangles. There are at least four distinct edges, each shared by a black triangle in B_i, $i = 1, 2$, and a white triangle incident to v. Each of these edges is a boundary edge and is incident to v, implying that the vertex v is a gap.

($\Leftarrow$) Let the object O have at least one gap v. There are at least two sets of contiguous (1-connected) black triangles around v, and at least two such sets of white triangles. Let B_1 and B_2 be such sets of black triangles. The set B_1 is surrounded on both sides with two white triangles t_1 and t_2 (B_1 contains a black triangle 1-adjacent to a white triangle t_1 in one of the sets of white triangles incident to v, and a black triangle 1-adjacent to a white triangle t_2 in the other set of white triangles incident to v).

- If B_1 and B_2 are 1-connected in O (if some black triangle in B_1 is 1-connected in O to some black triangle in B_2), then we can construct a polygonal Jordan curve J passing through the vertex v, and through the interior of all the black triangles in the sequence of 1-connected triangles connecting B_1 and B_2. Each such black triangle is 1-adjacent to the previous and the next triangle in the sequence through one of its three edges each (with the exception of the first and the last triangle in the sequence, which are edge-adjacent to the next and the previous triangle only, respectively, through one of the three incident edges).

  The curve J can be constructed by connecting the vertex v with the midpoints of these edges of the first and the last triangle in the sequence and by connecting the midpoints of the two "connecting" edges inside the other black triangles in the sequence, as illustrated in Fig. 2(a) for one of the configuration of black and white triangles around the gap v. One of the two (0-adjacent) white triangles t_1 and t_2 is inside J, the other is outside, and the two white triangles, being separated by the curve J lying (with the exception of the vertex v) completely in the interior of the object O, cannot be connected through a sequence of 1-adjacent white triangles. Thus, the triangles t_1 and t_2 belong to a 0-connected component of the complement O^c of O, which is not a 1-connected component.

- If B_1 and B_2, apart from being 0-adjacent through the vertex v, can be connected through a sequence of 0-adjacent black triangles not containing v (with at least one adjacency being strict), as illustrated in Fig. 2(b), or if B_1 and B_2 are not connected at all (except through v), as illustrated in Fig. 2(c), then the black triangles in B_1 and those in B_2 belong to a 0-connected component of O which is not a 1-connected component.

Thus, if the object O contains a gap, there is either a 0-connected component of O^c that is not a 1-connected component, or there is a 0-connected component of O that is not a 1-connected component.

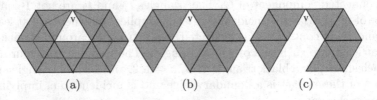

(a)   (b)   (c)

Fig. 2. Illustration of the ($\Leftarrow$) part of the proof of Proposition 3 when (a) B_1 and B_2 belong to a 1-connected component of O (the polygonal Jordan curve is drawn in red), (b) B_1 and B_2 do not belong to the same 1-connected component of O and a triangle in B_1 is connected to a triangle in B_2 through a sequence of pairwise 0-adjacent black triangles, with at least one adjacency being strict and (c) B_1 and B_2 are disconnected in O, with the exception of the 0-adjacency between a triangle in B_1 and a triangle in B_2 through the vertex v. (Color figure online)

5   The Number of Gaps in the Triangular Grid

We give a formula for the number of gaps in the triangular grid, as an extension of the known formula for the number of gaps in the square grid. It expresses the number of gaps in an object O through the numbers c_1^* and c_0^* of boundary edges and vertices in O, respectively.

Proposition 4. *The number of gaps in an object O in the triangular grid, with double gaps counted twice, is given by*

$$g^{(1)} + 2g^{(2)} = g + g^{(2)} = c_1^* - c_0^*.$$

Proof. We can prove this claim by induction on the number of triangles in O. For the base case, an object O consisting of a single triangle has three boundary edges, three boundary vertices and no gaps. Thus, the formula is valid for O.

Let us suppose the formula is valid for objects with n triangles, and let O be an object with $n+1$ triangles. We can assume, without loss of generality, that the grid is oriented so that it consists of ∇ and Δ triangles. Let t be the rightmost of the uppermost triangles in O and let $O' = O\backslash\{t\}$. For a Δ triangle t, there are 2^7 cases to consider, depending on the color (black or white) of the left and lower neighbors of t. For a ∇ triangle t, there are 2^4 such cases. For each case, we can show that the change in the number of gaps (taking into account their multiplicity) induced by adding the triangle t to O' is equal to the change in the difference between the number of boundary edges and vertices.

Instead of giving details on all the cases, we give three alternative proofs of this statement, inspired by the proofs of the analogous statement in the square grid [12,13]. The first proof is combinatorial, the second and the third one are based on graph theory.

First Proof: Note that c_0^* is the number of boundary vertices that may or may not be gaps. Thus,

$$c_0^* = \bar{g} + g$$

(recall that $\bar{g}$ denotes the number of non-gap boundary vertices).

Let us consider the set of all ordered pairs (e, v), where e is a boundary edge and v is a boundary vertex (non-gap or gap) incident to e. There are c_1^* cells e and $c_0^* = \bar{g} + g$ cells v.

Each boundary edge is incident to two boundary vertices. Each boundary non-gap vertex is incident to two, each simple gap is incident to four, and each double gap is incident to six boundary edges. Consequently, each boundary edge is in two, each boundary non-gap vertex is in two, each simple gap is in four and each double gap is in six such pairs (e, v). Thus

$$2c_1^* = 2\bar{g} + 4g^{(1)} + 6g^{(2)}$$
$$c_1^* = \bar{g} + 2g^{(1)} + 3g^{(2)}$$
$$c_1^* = \bar{g} + g^{(1)} + g^{(1)} + g^{(2)} + 2g^{(2)}$$
$$c_1^* = c_0^* + g^{(1)} + 2g^{(2)}$$

and

$$g^{(1)} + 2g^{(2)} = c_1^* - c_0^*$$
$$g + g^{(2)} = c_1^* - c_0^*.$$

Second Proof: Let us consider the graph $G = (N, A)$, where the nodes in N correspond to the boundary edges in O. Recall that each non-gap vertex is incident to two, each simple gap is incident to four and each double gap is incident to six boundary edges.

– For a non-gap vertex v, we connect the two nodes in N corresponding to the two boundary edges incident to v through an arc in A.

- For a simple gap v, the six triangles incident to v can be grouped into 1-connected sets of triangles of the same color (black or white). There are two such sets B_1 and B_2 of black triangles and two sets W_1 and W_2 of white ones. Two of the four boundary edges incident to v are both incident to triangles in W_1. We connect the two nodes in N corresponding to these two edges through an arc in A. The other two of the four boundary edges are incident to triangles in W_2. We connect the corresponding nodes in N through an arc in A.
- For a double gap v, there are three such sets of black and white triangles (each consisting of a single triangle). The six boundary edges incident to v can be grouped in three pairs, with the two edges in each pair being both incident to the same white triangle. We connect the nodes corresponding to the paired edges through an arc in A.

Thus, each non-gap boundary vertex generates one arc in A, each simple gap generates two arcs in A and each double gap generates three arcs. Each boundary edge is connected in the graph to exactly one edge across each of its two incident vertices.

There are c_1^* nodes in N and $\bar{g} + 2g^{(1)} + 3g^{(2)}$ arcs in A. The degree of each node in N is equal to two (the number of vertices in the boundary of the edge corresponding to the node). The sum of degrees of the nodes in N is equal to twice the number of arcs in A, i.e.,

$$2c_1^* = 2(\bar{g} + 2g^{(1)} + 3g^{(2)})$$
$$c_1^* = \bar{g} + g^{(1)} + g^{(1)} + g^{(2)} + 2g^{(2)}$$
$$c_1^* = c_0^* + g^{(1)} + 2g^{(2)}$$

and

$$g^{(1)} + 2g^{(2)} = c_1^* - c_0^*$$
$$g + g^{(2)} = c_1^* - c_0^*.$$

Third Proof: Let us consider the graph $G = (N, A)$, with nodes in N corresponding to boundary vertices in O, and arcs in A corresponding to boundary edges. Each node in N has an even degree: two if it corresponds to a non-gap boundary vertex, four if it corresponds to a simple gap and six if it corresponds to a double gap. Thus, there is an Eulerian cycle (containing all arcs in A exactly once) in the graph G. Let $v_1, e_1, \ldots, v_{c_1^*}, e_{c_1^*}$ be a cyclic ordering of nodes v_i and arcs e_i in C, $1 \leq i \leq c_1^*$. The number c_1^* of arcs in C is equal to the number of nodes in C. Non-gap boundary vertices appear once in C, simple gaps appear twice and double gaps appear three times. Thus

$$c_1^* = \bar{g} + 2g^{(1)} + 3g^{(2)}$$
$$c_1^* = \bar{g} + g^{(1)} + g^{(1)} + g^{(2)} + 2g^{(2)}$$
$$c_1^* = c_0^* + g^{(1)} + 2g^{(2)}$$

and

$$g^{(1)} + 2g^{(2)} = c_1^* - c_0^*$$
$$g + g^{(2)} = c_1^* - c_0^*.$$

6 Summary and Future Work

We defined gaps in binary objects in the triangular grid based on the number of incident boundary edges, i.e., as central vertices of certain configurations of the incident triangles, and we proposed a possible taxonomy for the gaps. We gave four definitions of well-composed objects in the triangular grid: based on the absence of gaps in O, based on the properties of the topological or combinatorial boundary of the object, and based on the influence of the chosen adjacency relation on the connectivity properties of the object, and we proved that the four definitions are equivalent. We gave a formula for the number of gaps (with double gaps counted twice), and three short and intuitive proofs of this formula.

Alternative formulas for the number of gaps in an object O in the triangular grid could be obtained, like in the square grid [10,13], expressing the number of gaps through the Euler characteristic of the object O. Another possible research direction could be an extension of this work to the 3D face centered cubic and the diamond grids.

Acknowledgement. We are grateful to the anonymous reviewers for careful reading of the paper and constructive comments. This work has been partially supported by the Ministry of Education and Science of the Republic of Serbia within the Project No. 34014.

References

1. Borgefors, G.: Distance transformations on hexagonal grids. Pattern Recognit. Lett. **9**(2), 97–105 (1989)
2. Borgefors, G., Sanniti di Baja, G.: Skeletonizing the distance transform on the hexagonal grid. In: 9th International Conference on Pattern Recognition, ICPR, pp. 504–507 (1988)
3. Boutry, N.: A study of well-composedness in n-D (Une étude du bien-composé en dimension n). Ph.D. thesis, University of Paris-Est, France (2016)
4. Boutry, N., Géraud, T., Najman, L.: How to make nD images well-composed without interpolation. In: 2015 IEEE International Conference on Image Processing, ICIP 2015, pp. 2149–2153 (2015)
5. Boutry, N., Géraud, T., Najman, L.: A tutorial on well-composedness. J. Math. Imaging Vis. **60**(3), 443–478 (2018)
6. Bribiesca, E.: A new chain code. Pattern Recognit. **32**(2), 235–251 (1999)
7. Brimkov, V.E.: Formulas for the number of $(n-2)$-gaps of binary objects in arbitrary dimension. Discret. Appl. Math. **157**(3), 452–463 (2009)
8. Brimkov, V.E., Maimone, A., Nordo, G.: An explicit formula for the number of tunnels in digital objects. CoRR abs/cs/0505084 (2005). http://arxiv.org/abs/cs/0505084
9. Brimkov, V.E., Maimone, A., Nordo, G.: Counting gaps in binary pictures. In: Reulke, R., Eckardt, U., Flach, B., Knauer, U., Polthier, K. (eds.) IWCIA 2006. LNCS, vol. 4040, pp. 16–24. Springer, Heidelberg (2006). https://doi.org/10.1007/11774938_2

10. Brimkov, V.E., Maimone, A., Nordo, G., Barneva, R.P., Klette, R.: The number of gaps in binary pictures. In: Bebis, G., Boyle, R., Koracin, D., Parvin, B. (eds.) ISVC 2005. LNCS, vol. 3804, pp. 35–42. Springer, Heidelberg (2005). https://doi.org/10.1007/11595755_5

11. Brimkov, V.E., Moroni, D., Barneva, R.: Combinatorial relations for digital pictures. In: Kuba, A., Nyúl, L.G., Palágyi, K. (eds.) DGCI 2006. LNCS, vol. 4245, pp. 189–198. Springer, Heidelberg (2006). https://doi.org/10.1007/11907350_16

12. Brimkov, V.E., Nordo, G., Barneva, R.P., Maimone, A.: Genus and dimension of digital images and their time- and space-efficient computation. Int. J. Shape Model. 14(2), 147–168 (2008)

13. Čomić, L.: On gaps in digital objects. In: Barneva, R., Brimkov, V., Tavares, J. (eds.) IWCIA 2018. LNCS, vol. 11255, pp. 3–16. Springer, Cham (2018). https://doi.org/10.1007/978-3-030-05288-1_1

14. Čomić, L., Magillo, P.: Repairing 3D binary images using the BCC grid with a 4-valued combinatorial coordinate system. Inf. Sci. (2019, to appear)

15. Deutsch, E.S.: On parallel operations on hexagonal arrays. IEEE Trans. Comput. 19(10), 982–983 (1970)

16. Deutsch, E.S.: Thinning algorithms on rectangular, hexagonal, and triangular arrays. Commun. ACM 15(9), 827–837 (1972)

17. Dutt, M., Andres, E., Largeteau-Skapin, G.: Characterization and generation of straight line segments on triangular cell grid. Pattern Recognit. Lett. 103, 68–74 (2018)

18. Dutt, M., Biswas, A., Nagy, B.: Number of shortest paths in triangular grid for 1- and 2-neighborhoods. In: Barneva, R.P., Bhattacharya, B.B., Brimkov, V.E. (eds.) IWCIA 2015. LNCS, vol. 9448, pp. 115–124. Springer, Cham (2015). https://doi.org/10.1007/978-3-319-26145-4_9

19. Evako, A.V., Kopperman, R., Mukhin, Y.V.: Dimensional properties of graphs and digital spaces. J. Math. Imaging Vis. 6(2–3), 109–119 (1996)

20. Freeman, H.: Algorithm for generating a digital straight line on a triangular grid. IEEE Trans. Comput. 28(2), 150–152 (1979)

21. Golay, M.J.E.: Hexagonal parallel pattern transformations. IEEE Trans. Comput. 18(8), 733–740 (1969)

22. González-Díaz, R., Jiménez, M.-J., Medrano, B.: 3D well-composed polyhedral complexes. Discret. Appl. Math. 183, 59–77 (2015)

23. González-Díaz, R., Jiménez, M.-J., Medrano, B.: Efficiently storing well-composed polyhedral complexes computed over 3D binary images. J. Math. Imaging Vis. 59(1), 106–122 (2017)

24. Gray, S.: Local properties of binary images in two dimensions. IEEE Trans. Comput. 20, 551–561 (1971)

25. Her, I.: Geometric transformations on the hexagonal grid. IEEE Trans. Image Process. 4(9), 1213–1222 (1995)

26. Kardos, P., Palágyi, K.: Topology preservation on the triangular grid. Ann. Math. Artif. Intell. 75(1–2), 53–68 (2015)

27. Kardos, P., Palágyi, K.: On topology preservation of mixed operators in triangular, square, and hexagonal grids. Discret. Appl. Math. 216, 441–448 (2017)

28. Klette, R., Rosenfeld, A.: Digital Geometry: Geometric Methods for Digital Picture Analysis. Morgan Kaufmann Publishers, San Francisco, Amsterdam (2004)

29. Kong, T.Y., Rosenfeld, A.: Digital topology: introduction and survey. Comput. Vis., Graph., Image Process. 48(3), 357–393 (1989)

30. Kovalevsky, V.A.: Geometry of Locally Finite Spaces: Computer Agreeable Topology and Algorithms for Computer Imagery. Editing House Dr. Bärbel Kovalevski, Berlin (2008)

31. Latecki, L.J.: 3D well-composed pictures. CVGIP: Graph. Model Image Process. **59**(3), 164–172 (1997)

32. Latecki, L.J., Eckhardt, U., Rosenfeld, A.: Well-composed sets. Comput. Vis. Image Underst. **61**(1), 70–83 (1995)

33. Maimone, A., Nordo, G.: On 1-gaps in 3d digital objects. Filomat **22**(3), 85–91 (2011)

34. Maimone, A., Nordo, G.: A formula for the number of $(n − 2)$-gaps in digital n-objects. Filomat **27**(4), 547–557 (2013)

35. Nagy, B.: Cellular topology and topological coordinate systems on the hexagonal and on the triangular grids. Ann. Math. Artif. Intell. **75**(1–2), 117–134 (2015)

36. Nagy, B., Lukic, T.: Dense projection tomography on the triangular tiling. Fundam. Inform. **145**(2), 125–141 (2016)

37. Nagy, B., Strand, R.: Approximating Euclidean circles by neighbourhood sequences in a hexagonal grid. Theor. Comput. Sci. **412**(15), 1364–1377 (2011)

38. Sarkar, A., Biswas, A., Dutt, M., Bhowmick, P., Bhattacharya, B.B.: A linear-time algorithm to compute the triangular hull of a digital object. Discret. Appl. Math. **216**, 408–423 (2017)

39. Sarkar, A., Biswas, A., Dutt, M., Mondal, S.: Finding shortest triangular path and its family inside a digital object. Fundam. Inform. **159**(3), 297–325 (2018)

40. Siqueira, M., Latecki, L.J., Tustison, N.J., Gallier, J.H., Gee, J.C.: Topological repairing of 3D digital images. J. Math. Imaging Vis. **30**(3), 249–274 (2008)

41. Sossa-Azuela, J.H., Cuevas-Jiménez, E.V., Zaldivar-Navarro, D.: Computation of the Euler number of a binary image composed of hexagonal cells. J. Appl. Res. Technol. **8**, 340–350 (2010)

42. Stelldinger, P., Latecki, L.J., Siqueira, M.: Topological equivalence between a 3D object and the reconstruction of its digital image. IEEE Trans. Pattern Anal. Mach. Intell. **29**(1), 126–140 (2007)

43. Wiederhold, P., Morales, S.: Thinning on quadratic, triangular, and hexagonal cell complexes. In: Brimkov, V.E., Barneva, R.P., Hauptman, H.A. (eds.) IWCIA 2008. LNCS, vol. 4958, pp. 13–25. Springer, Heidelberg (2008). https://doi.org/10.1007/978-3-540-78275-9_2

Generating Second Order (Co)homological Information within AT-Model Context

Pedro Real[1], Helena Molina-Abril[1(✉)], Fernando Díaz del Río[1], and Darian Onchis[2,3]

[1] H.T.S. Informatics' Engineering, University of Seville, Seville, Spain
{real,habril}@us.es, fdiaz@atc.us.es
[2] Faculty of Mathematics, University of Vienna, Vienna, Austria
darian.onchis@univie.ac.at
[3] Faculty of Mathematics and Computer Science, West University of Timisoara, Timişoara, Romania

Abstract. In this paper we design a new family of relations between (co)homology classes, working with coefficients in a field and starting from an AT-model (Algebraic Topological Model) $AT(C)$ of a finite cell complex C These relations are induced by elementary relations of type "to be in the (co)boundary of" between cells. This high-order connectivity information is embedded into a graph-based representation model, called *Second Order AT-Region-Incidence Graph (or AT-RIG)* of C. This graph, having as nodes the different homology classes of C, is in turn, computed from two generalized abstract cell complexes, called *primal and dual AT-segmentations* of C. The respective cells of these two complexes are connected regions (set of cells) of the original cell complex C, which are specified by the integral operator of $AT(C)$. In this work in progress, we successfully use this model (a) in experiments for discriminating topologically different 3D digital objects, having the same Euler characteristic and (b) in designing a parallel algorithm for computing potentially significant (co)homological information of 3D digital objects.

Keywords: Cell complex · Algebraic-topological model
Homology computation · Primal and dual AT-segmentation
AT-model region-incidence-graph · nD digital object

1 Introduction

(Co)homology (see for instance [33]) provides valuable information about topological spaces, by observing sets that intuitively have no (co)boundary, but are

This work has been supported by the Spanish research projects MTM2016-81030-P (AEI/FEDER, UE) and TEC2012-37868-C04-02, and by the VPPI of the University of Seville. Darian Onchis gratefully acknowledges the support of the Austrian Science Fund FWF-P27516.

© Springer Nature Switzerland AG 2019
R. Marfil et al. (Eds.): CTIC 2019, LNCS 11382, pp. 68–81, 2019.
https://doi.org/10.1007/978-3-030-10828-1_6

on the (co)boundary of other sets. These sets are representative (co)cycles of a (co)homology hole, seen as an equivalence class. Algebraic (co)homological information with coefficients in a field could be defined as the set of processed and structured linear algebraic data describing in some sense its (co)homology classes and the relations between them. We talk about homology and cohomology information as a whole due to the fact that homology and cohomology classes are measured using different strategies (delineating or cutting holes) for detecting homological holes over the initial topological data. A simple example of (co)homology information is provided by the numerical topological invariants called Betti numbers. For instance, if X is a cell complex embedded in $\mathbb{R}^3$, Betti numbers β_0, β_1 and β_2 respectively measure the number of different connected components, tunnels and cavities of X.

Roughly speaking, (co)homotopy holes of objects (those related to delineating or cutting generalized "parametrized and oriented closed curves") are theoretically attainable from homology's ones [23], but these methods have an enormous complexity in time and space [4]. An easier relation between homology and homotopy is given by the Euler characteristic (see [1]), defined in local terms as the alternate sum of the number of cells in each dimension. This number is the most simple example of homotopy invariant, that can also be obtained from the global homological information provided by the Betti numbers.

Now, (co)homology information of X is not reduced in general to that provided by Betti numbers. For example, a torus T and a three-dimensional sphere with two handles S have the same Betti numbers (and, consequently the same Euler characteristic) but they are not (co)homologically equivalents. The two tunnels of T are related to its cavity in a much more "stronger" way that the tunnels of S are with regards to the corresponding cavity.

We progress here in discovering the homotopy nature of homology, by creating two (non-unique) abstract cell complexes, called *primal and dual AT-segmentations*, both with significantly smaller number of cells than the original geometric cell complex C and from which it is possible to detect topological relationships between (co)homology classes of C with coefficient in a field $\mathbb{F}$. We construct the primal and dual segmentation with the help of an algebraic-topological model $AT(C)$ of C (or AT-model for short) [17,18,34,35]. Using the bounding functions of a primal and dual AT-segmentations and the relationship between cells "to be in the boundary of", we are finally able to compute a graph-based model $\mathcal{P}(AT(C))$, called *AT-model Region-Incidence-Graph* (or, *AT-RIG*, for short), whose nodes are the different homology classes of C.

We successfully use this technique in a set of experiments for discriminating topologically different 3D digital objects with the same Euler characteristic. We also use this modus operandi in designing a parallel algorithm for computing potential high-order homology statistics for a 3D digital objects. In a near future, we intend to study the corresponding degrees of independence with regards the AT-model chosen and of homology and homotopy invariances of an AT-RIG.

1.1 Related Works

Focusing on homotopy representation models of digital objects and images, there are numerous works that arise from sources of digital topology [2,21,25], continuous or cellular topology [9,20,26] and nD shape search with three clearly differentiated notions: Reeb graphs [5,13], skeletons [6,36,41] and boundary representations [3,16,28]. Relative to the intermediary step of homological computation of cell complexes, there is plenty of literature based on a pure algebraic perspective devoted to this issue. The classical method is based on the diagonalization of cell-incidence matrices to Smith normal form (SNF) [33]. Some advances in the computation of the SNF have been achieved [10], but the most successful approaches consist of reducing the number of cells in the complex using discrete-vector-field dynamics (Discrete Morse theory [12]) before computing the SNF for the small resulting cell complex (see, for instance, [7,14,19,22,32,35,39,40]). This paper goes beyond homological computation and designs (sequential and parallel) algorithms for computing a new graph-based representation that allows to discriminate homologically different geometric objects embedded in $\mathbb{R}^n$ having the same Betti numbers. In this sense, AT-segmentation theory extend and greatly improve both the algebraic model called Algebraic-Topological model [17,18,34,35,37] and the combinatorial model called Homological Spanning Forest (HSF, for short) [8,30,31,38] in this search of topological representations within digital image context.

2 Cell Complexes and Algebraic-Topological Models

We work in this paper with cell complex representations (composed of cells and bounding relations between them), that allow to model, for example, not only an n-dimensional digital object at sub-n-xel level but also significant algebraic (co)homological information (with coefficient in a field $\mathbb{F}$).

First at all, we provide a slightly modified version of the classical abstract cell complex notion (see [24] for a survey).

We say that $\mathsf{C} = (C, B, dms)$ is an *abstract cell complex (or ACC, for short)* if:

- $C = \{C_q\}_{q \in \mathbb{N} \cup \{0\}}$ is a finite set with a gradation $dms : C \to \mathbb{N} \bigcup \{0\}$ defined by $dms(c) = q$ for $c \in C_q$;
- $B : C \times C \to \mathbb{N} \cup \{0\}$ is a map such that satisfies the following condition: $B(c, c') \neq 0$ implies $c \in C_{q-1}$, $c' \in C_q$.

We refer to the elements of C as cells and to $B(c, c')$ as the bounding function of the ACC C applied to the couple (c, c'). If we extend the bounding function of the ACC in an antisymmetric and transitively way, we recover the classical notion of ACC.

The *connectivity-graph* $G(C, B, dms) = (V, E)$ of an abstract cell complex (C, B, dms) is the graph whose set of nodes is C and an edge $\{c, c'\} \in E$ if $B(c, c')$ or $B(c', c)$ is different from zero.

Now, let us define the (algebraic) notion of geometric cell complex. We say that $\mathsf{C} = (C, \kappa, dms)$ is a *Lefschetz complex* [27] if:

- $C = \{C_q\}_{q \in \mathbb{N} \cup \{0\}}$ is a finite set with a gradation $dms : C \to \mathbb{N} \cup \{0\}$ defined by $dms(c) = q$ for $c \in C_q$;
- $\kappa : C \times C \to \mathbb{F}$ is a map such that $\kappa(c, c') \neq 0$ implies $c \in C_{q-1}$, $c' \in C_q$. For any $c, c'' \in C$ we have $\sum_{c' \in C} \kappa(c, c')\kappa(c', c'') = 0$.

We refer to the elements of C as cells and to $\kappa(x, y)$ as the incidence coefficient of x, y.

In fact, an equivalent definition of a Lefschetz complex is that of a free chain complex $(\mathbb{F}[C], \partial_C)$ with *boundary* $\partial_C : \mathbb{F}[C] \to \mathbb{F}[C]$ defined on generators by $\partial_C(c) = \sum_{c' \in C} \kappa(c', c)c'$. Its *coboundary* $\delta_C : \mathbb{F}[C] \to \mathbb{F}[C]$ is defined on generators by $\delta_C(c) = \sum_{c' \in C} \kappa(c, c')c'$. The Lefschetz homology (resp. cohomology) of (C, κ, dms), denoted $H(C, \partial_C)$ (resp. $H(C, \delta_C)$) is the homology of the chain complex $(\mathbb{F}[C], \partial)$ (resp. $(\mathbb{F}[C], \delta)$). We are interested here in Lefschetz complexes satisfying that for any $c, c' \in C$ the incidence coefficient $\kappa(c, c')$ is either zero or ± 1 of $\mathbb{F}$. These structures are simply called here *geometric cell complexes*. The identity function $1_C : C \to C$ is defined by $1_C(c) = c$, $\forall c \in C$. Associated to a geometric cell complex (C, κ, dms), there is a bounding function $B : C \times C \to \mathbb{N} \cup \{0\}$ defined by $B(c, c') = 1$ if $\kappa(c, c') \neq 0$ and $B(c, c') = 0$ in the rest of cases. If $c \in C$ and R is a set of cells of (C, κ, dms), we define the bounding function $B(c, R) = \sum_{c' \in R} B(c, c')$ (resp. $B(R, c) = \sum_{c' \in R} B(c, c')$).

From now on, we use the triplet (C, ∂, dms) for denoting a geometric cell complex. We use the notation $c' \in f$, being $f \in \mathbb{F}[C]$, for indicating that the cell c' is involved as a non-null summand in this linear combination.

It is straightforward to specify geometric cell complexes structures modeling n-dimensional digital images at sub-n-xel level. In fact, we are mainly interested in running the designed algorithms for "cellularizations" of digital objects and images, in order to progress in topological acuity and representation within digital image context.

Now, we are able to define an *algebraic-topological model* (or AT-model, for short) (C, ∂, ϕ, dms) of a geometric cell complex (C, ∂, dms) (see [17]). The homomorphism $\phi : \mathbb{F}[C_q] \to \mathbb{F}[C_{q+1}]$, called *integral operator*, satisfies the following three conditions: (a) $\phi\phi = 0$; (b) $\partial\phi\partial = \partial$; (c) $\phi\partial\phi = \phi$. From this data, we can construct an explicit homology equivalence between the chain complex $(\mathbb{F}[C], \partial)$ and a free chain complex with null differential (which, obviously, is isomorphic to the homology $H(C, \partial)$). The germ idea of the AT-model comes back to the original notion of chain contraction exhaustively used by Samuel Eilenberg and Saunder Mac Lane in their works of homological computation (see, for example, [11]) in the fifties of the twentieth century.

3   AT-Segmentations

In this section and with the help of an AT-model (C, ∂, ϕ, dms), we construct two special partitions of C into connected regions from which it is possible to compute "strong" topological relations between homology classes, derived from the elementary relation "to be in the boundary of". From now on, we work with $\mathbb{F} = \mathbb{Z}_2$, in order to avoid the use of signs in the AT-model construction.

Let us emphasize that all the study done here can be correctly developed for any ground field.

Some terminology relative to primal regions of the primal AT-segmentation we want to construct is necessary. The dimension of a primal region R composed by cells of C of dimension t $(0 \leq t \leq n - 1)$ and, possibly, of dimension $t + 1$ is $dim^{pr}(R) = (t, t + 1)$. Its criticality number $crt(R)$ is given by the difference between the number of t-cells and $(t + 1)$-cells. All the regions of a primal AT-segmentation have criticality number greater or equal to zero. A primal region R with $crt(R) > 0$ is called *homologically essential*. If $crt(R) = 0$ is called *homologically inessential*.

Algorithm 1 uses as input a filtration of the geometric cell complex C. It is possible to design an algorithm for computing a primal AT-segmentation independent of this restriction. Such algorithm is based on the construction of hierarchical "spanning forests" within the global connectivity graph of C as ACC. There is no space here to address this question in detail. The reference [37] can be of help in the planning of such algorithm.

Figure 1 shows a primal AT-segmentation over an ACC version $Cell(O)$ of a 2D digital object.

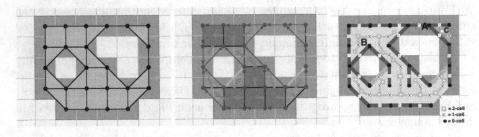

Fig. 1. (Left) ROI consisting of the set of black pixels. The implicit cellularization of the ROI -using 8-adjacency and being the 0-cells the square physical pixels- is superimposed. (Center) Vectors of cracks involved in the AT-model construction of Algorithm 1 are highlighted. (Right) The associated AT-segmentation of the ROI. There are three regions. Region A is drawn in red and is composed by all the 24 (N_0^A) 0-cells and 23 (N_1^A) 1-cells (a subdivision of the spanning tree of the 0-cells). Region B is drawn in yellow and is composed by 18 (N_1^B) 1-cells and 16 (N_2^B) 2-cells. Region C is drawn in green and its tree has one 1-cell and one 2-cell (thus, it is inessential). Regions A and B are homologically essential, due to the fact that $crt(R) = N_i^R - N_{i+1}^R = 1 > 0$, for $(R, i) = \{(A, 0), (B, 1)\}$. In fact, A detects one 0-dimensional homology class and B two 1-dimensional holes. (Color figure online)

Let us note that the connectivity graph of regions of the primal AT-segmentation $HS^{pr}(C)$ is not necessarily a tree. For each critical cell $e_q^{d_q}$, $(1 \leq q \leq m)$ its corresponding primal segmentation region $\mathcal{S}_\ell^{pr}(e_q^{d_q})$ of dimension $(d_q, d_q + 1)$ has a criticality number $crt(\mathcal{S}_\ell^{pr}(e_q^{d_q}))$ greater than zero. The rest of the primal segmentation regions have a criticality number equal to zero.

Algorithm 1. [primal AT-segmentation]

Input: A geometric cell complex $C := \{C, \partial, dms\}$. C is a list with all the cells of C ordered by increasing dimension $c_1^0, \ldots c_{\ell_1}^0, c_1^1, \ldots, c_{\ell_2}^1, \ldots, c_1^n, \ldots, c_{\ell_n}^n$. Here, $dms(c_j^k) = k, \forall k, j$ and $\sum_{1 \leq q \leq n} \ell_q = \ell$. Let us also use the cell ordering $c_j^k = c_{j + \sum_{q < k} \ell_q}$. The boundary operator $\partial|_{\mathbb{F}[c_1, \ldots, c_i]}$ is denoted by ∂_i.

1:   $\mathcal{H}_0^\partial \leftarrow \emptyset; \mathcal{K}_0^\phi \leftarrow \emptyset; \mathcal{J}_0^\phi \leftarrow \emptyset; \mathcal{S}_0^{pr} \leftarrow \emptyset;$

2:   **for** $k = 0$ to n **do**

3:      **for** $j = 1$ to ℓ_k **do**

4:         $i \leftarrow j + \sum_{q < k} \ell_q;$

5:         $\phi_{i-1}(c_i) \leftarrow 0;$

6:         $\mathcal{R}_i \leftarrow \{c_i\};$

7:         $\mathcal{S}_i^{pr} \leftarrow \mathcal{S}_{i-1}^{pr} \cup \{\mathcal{R}_i\}$. Let us denote by $\mathcal{S}_i^{pr}(c_q)$ the region of $\mathcal{S}_i^{pr}$ containing the cell c_q $(1 \leq q \leq i)$. In this way, $\mathcal{S}_i^{pr}$ is handled as the union $\bigcup_{1 \leq q \leq i} \{\mathcal{S}_i^{pr}(c_q)\}$.

8:         $Bnd_i \leftarrow \{e \in \partial_i(c_i)\};$ $\triangleright$ Boundary of the current cell

9:         $\bar{c}_i \leftarrow c_i + \phi_{i-1}\partial_i(c_i);$ $\triangleright$ Potential cycle associated to c_i

10:        $\overline{Bnd}_i \leftarrow \{e \in \partial_i(\bar{c}_i)\};$ $\triangleright$ Algebraic boundary of $\bar{c}_i$

11:        $\widetilde{Bnd}_i \leftarrow \{e \in (1_C + \partial_i\phi_{i-1})(c), \text{ for some } c \in Bnd_i\};$ $\triangleright$ Combinatorial boundary of $\bar{c}_i$

12:        $\mathcal{N}_i \leftarrow \{e \in Bnd_i : dim^{pr}(\mathcal{S}_i^{pr}(e)) = (dms(c_i) - 1, dms(c_i)) \wedge crt(\mathcal{S}_i^{pr}(e)) = 0\};$

        $\triangleright$ Homologically inessential regions in the boundary of c_i

13:        $\mathcal{H}_i^\partial \leftarrow \mathcal{H}_{i-1}^\partial \bigcup \{\bar{c}_i\}; \mathcal{K}_i^\phi \leftarrow \mathcal{K}_{i-1}^\phi \bigcup \{c_i\}; \mathcal{J}_i^\phi \leftarrow \mathcal{J}_{i-1}^\phi \bigcup \{c_i\};$ $\triangleright$ Homology generators, combinatorial kernel, critical cells

14:        **if** $\partial_i(\bar{c}_i) == 0$ **then** $\triangleright$ Equivalent to $\overline{Bnd}_i == \emptyset$

15:           **for** $r = 1$ to i **do**

16:              $\phi_i(c_r) \leftarrow \phi_{i-1}(c_r);$

17:        **else** $\triangleright$ In case in which c_i does not generate a cycle

18:           $\overline{\mathcal{J}}_i \leftarrow \overline{Bnd}_i \bigcap \mathcal{J}_i^\phi$

19:           $\tilde{\mathcal{J}}_i \leftarrow \widetilde{Bnd}_i \bigcap \mathcal{J}_i^\phi$

20:           $\mathcal{S}_i^{pr}(c_i) \leftarrow \mathcal{S}_i^{pr}(c_i) \bigcup_{e \in \tilde{\mathcal{J}}_i \cup \mathcal{N}_i} \mathcal{S}_i^{pr}(e);$ $\triangleright$ Updating the primal partition

21:           **for** $e \in \tilde{\mathcal{J}}_i \cup \mathcal{N}_i$ **do**

22:              $\mathcal{S}_i^{pr}(e) \leftarrow \mathcal{S}_i^{pr}(c_i);$

23:           Choose one of the cells $\mathbf{e} \in \overline{\mathcal{J}}_i$ $\triangleright$ Updating AT-model

24:           $\tilde{\phi}(\mathbf{e}) \leftarrow c_i;$

25:           $\tilde{\phi}(c) \leftarrow 0$ for each $c \in \mathcal{K}_{i-1}^\phi \setminus \{\mathbf{e}\};$

26:           $\bar{\mathbf{e}} \leftarrow \mathbf{e} + \phi_{i-1}\partial_i(\mathbf{e});$

27:           **for** $q = 1$ to $i - 1$ **do**

28:              $\phi_i(c_q) \leftarrow (\phi_{i-1} + (id_{C_i} - \phi_{i-1}\partial_{i-1})\tilde{\phi}(id_{C_i} - \partial_{i-1}\phi_{i-1}))(c_q), \forall c_q \in C_i$

29:           $\mathcal{H}_i^\partial \leftarrow \mathcal{H}_i^\partial \setminus \{\bar{\mathbf{e}}, \bar{c}_i\};$ $\triangleright$ Updating homology generators

30:           $\mathcal{K}_i^\phi \leftarrow \mathcal{K}_i^\phi \setminus \{\mathbf{e}\};$ $\triangleright$ Updating combinatorial homology kernel

31:           $\mathcal{J}_i^\phi \leftarrow \mathcal{J}_i^\phi \setminus \{\mathbf{e}, c_i\};$ $\triangleright$ Updating set of critical cells

32: $B^{pr}[C](R, R') \leftarrow \#\{c' \in R' : dms(c') = t \wedge B(R, c') = \sum_{c \in R} B(c, c') \neq 0\}$

     $\forall R, R' \in \mathcal{S}_\ell^{pr}$, with $dim^{pr}(R) = (t-1, t)$ and $dim^{pr}(R') = (t, t+1)$ $(1 \leq t \leq n)$.

     $\triangleright$ Specifying primal AT-segmentation bounding function

Output:

- An AT-model $(C, \partial_\ell, \phi_\ell, dms)$ and a combinatorial basis (set of *critical cells*) specified by $\mathcal{J}_\ell^\phi$ and ordered by increasing dimension $\{e_1^{d_1}, \ldots, e_m^{d_m}\}$ (with $dms(e_q^{d_q}) = d_q, 1 \leq q \leq m$) for the homology $H(C, \phi)$ given by $\mathcal{H}_\ell^\phi$.

- An abstract cell complex $HS^{pr}(C)$, called *primal AT-segmentation of* C, whose set of cells is the partition in regions $\mathcal{S}_\ell^{pr}$ of C and its bounding function is $B^{pr}[C]$.

From this output, it is possible to define a dual AT-segmentation $HS^{dl}(\mathsf{C})$ in this simple manner:

- [Initial dual AT-partition]. We consider as initial dual AT-partition $\overline{\mathcal{S}}_{\ell}^{dl}$ a refinement of the primal segmentation partition $\mathcal{S}_{\ell}^{pr}$, in which each critical cell (as sets formed by one element) is considered as a new region of the partition. Let us note that all the regions of $\overline{\mathcal{S}}_{\ell}^{dl}$ have now zero as criticality number, excepting the sets formed by one critical cell, which have one as criticality number.

- [Updating initial dual AT-partition]. For each critical cell $e_q^{d_q} \in \mathcal{H}_{\ell}^{\phi}$ $(1 \leq q \leq m)$, let us construct the region of the dual segmentation $\overline{\mathcal{S}}_{\ell}^{dl}(e_q^{d_q}) = \{e_q^{d_q}\} \bigcup_{c \in \partial(e_q^{d_q})} \overline{\mathcal{S}}_{\ell}^{pr}(c)$ of dimension $dim^{dl}(\mathcal{S}_{\ell}^{dl}(e_q^{d_q})) = (d_q - 1, d_q)$. After updating the regions of the partition corresponding to the critical cells, the rest of regions of $\overline{\mathcal{S}}_{\ell}^{dl}$ remain unaltered. The resulting partition describes at set level the desired dual AT-segmentation and it is denoted by $\mathcal{S}_{\ell}^{dl}$. Let us emphasize that the ranks (that is, the difference between the number of cells of C and the number of regions of the partition) of the primal and dual AT-partitions can be different. This is mainly due to the fact that, in general, there is no one-to-one relation between critical cells and regions of the primal or dual AT-segmentations.

- [Dimension and Bounding Function]. The dimension of a region $R \in \mathcal{S}_{\ell}^{dl}$, having t-cells and, possibly $t - 1$ cells is $dim^{dl}(R) = (t, t - 1)$. Its bounding function is defined by $B^{dl}[\mathsf{C}](R, R') = \#\{c \in R : dms(c) = t \wedge B(c, R') = \sum_{c' \in R'} B(c, c') \neq 0\}$ $\forall R, R' \in \mathcal{S}_{\ell}^{dl}$, with $dim^{dl}(R) = (t, t - 1)$ and $dim^{dl}(R') = (t + 1, t)$ $(1 \leq t \leq n)$.

At the end of this process, we get a dual AT-segmentation $HS^{dl}(\mathsf{C}) = (\mathcal{S}_{\ell}^{dl}, B^{dl}[\mathsf{C}], dim^{dl})$.

Let us note that the connectivity graph of regions of $HS^{dl}(\mathsf{C})$ is not necessarily a tree.

We are now ready to build a second order AT-RIG associated with the primal and dual AT-segmentations. AT-RIGs measure all the relationships between two homology classes of dimension t and $t+1$, $\forall 0 \leq t \leq n$. Let us write in pseudocode the construction of the second order primal AT-RIG (see Algorithm 2). The construction of the dual one is completely analogous.

Finally, the second order AT-RIG $G_1(AT(\mathsf{C}))$ is the graph whose nodes are the different homology classes (represented by their corresponding critical cell) and whose edges are those belonging to both primal and dual AT-RIGs.

4   Operations with AT-Segmentations

Given a primal (or dual) AT-segmentation, it is possible to create a new one changing only the participation of two cells. This operation is called *crack transport* and is exhaustively used in the parallel methods for computing homology information designed in [38].

Algorithm 2. [Second order Primal AT-RIG]

Input:

- an AT-model $(C, \partial_\ell, \phi_\ell, dms)$ and a combinatorial basis (set of *critical cells*) ordered by increasing dimension $\{e_1^{d_1}, \ldots, e_m^{d_m}\}$ (with $dms(e_q^{d_q}) = d_q$, $1 \leq q \leq m$) for the homology $H(C, \phi)$.
- a primal AT-segmentation $HS^{pr}(C) = (S_\ell^{pr}, B^{pr}[C], dim^{pr})$

```
1:  for i = 1 to m do
2:      v_i ← e_i^{d_i};                        ▷ the nodes of the RIG are the critical cells
3:      N(v_i) ← ∅          ▷ The set of all the neighbors of the region of the primal
    AT-segmentation containing v_i
4:      N_s(v_i) ← ∅              ▷ the set of neighbors of v_i in the second order AT-RIG
5:      for j = 1 to i do
6:          if B^{pr}[C](S_ℓ^{pr}(v_j), S_ℓ^{pr}(v_i)) ≠ 0 then
7:              N(v_i) ← N(v_i) ∪ {v_j};
8:              if B^{pr}[C](S_ℓ^{pr}(v_j), S_ℓ^{pr}(v_i)) = #{d_i − cells ∈ S_ℓ^{pr}(v_i)} then
9:                  N_s(v_i) ← N_s(v_i) ∪ {v_j};
10:                 B_{rig}^{pr}[C](v_j, v_i) ← 1;
11:         else
12:             B_{rig}^{pr}[C](v_j, v_i) ← 0
```
Output: The region-incidence-graph $G_1^{pr}(AT(C))$ associated to the abstract cell complex $(\{e_1^{d_1}, \ldots, e_m^{d_m}\}, B_{rig}^{pr}[C], dms)$.

Algorithm 3 shows the admissible Crack Transport Algorithm. Crack transports can be used for AT-segmentation parallel computation.

Algorithm 3. [Admissible Crack Transport Algorithm].

Let $HS(C)$ be a primal AT-segmentation of a geometric cell complex C. Let R_1 and R_2 be two regions of dimension $(k-1, k)$ and $(k, k+1)$ respectively, and $c \in R_1$ and $c' \in R_2$ be two k-cells. Let us denote by $\tilde{HS}(C)$ the segmentation $HS \setminus \{c, c'\}$. The new segmentation $[c \leftrightarrow c']HS(C)$ resulting from the initial one, assigning c to R_2 and c' to R_1 is a new primal AT-segmentation if (a) c is incident to R_2, c' is incident to R_1;(b) there is at least a pair (S_1, S_2) of regions of $\tilde{HS}(C)$ with $S_1 \subset R_1$ $(dim^{pr}(S_1) = (k-1, k))$ and $S_2 \subset R_2$ $(dim^{pr}(S-2) = (k, k+1))$, satisfying that:

- $\#(S_1 \cap \partial c) = 1$ and $\#(S_1 \cap \partial c') = 1$;
- $\#(S_2 \cap \delta c) = 1$ and $\#(S_2 \cap \delta c') = 1$;
- $B^{pr}[C](S_1, S_2) > 1$.

In Fig. 2, an internal (within the ROI) crack transport defined as an admissible interchange of cells between "connected" homological regions of the AT-segmentation of Fig. 1 is shown.

Another example of application of crack transport is shown in Fig. 3.

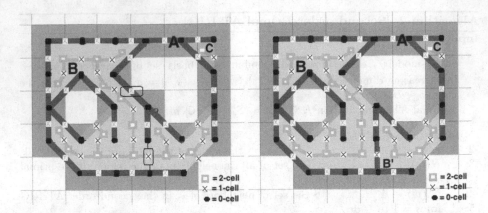

Fig. 2. (Left) AT-segmentation of Fig. 1. 1-cells involved in the internal crack transport are surrounded by black closed curves; (Right) Result of the crack transport. There are now three 1–2 regions B, B', C being the first two ones homologically essentials ($\beta^B = 1 = \beta^{B'}$). Both AT-segmentations present the same second order AT-RIG: a tree with three hole-nodes ($\alpha^0, \beta^1, \gamma^1$), connecting α^0 with β^1 and γ^1.

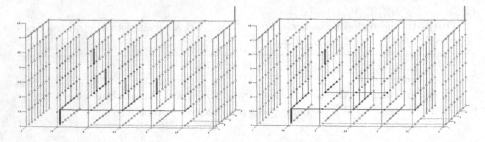

Fig. 3. A ROI composed of three segments parallel to the axis (a total of 5 black voxels). Points represent 0-cells (voxels), triangles 1-cells, squares 2-cells and stars 3-cells. Two AT-segmentations of the whole image that embeds the ROI are drawn. The AT-segmentation on the right is the result of several crack transport operations on the AT-segmentation on the left. Obviously, both have the same trivial AT-RIG: a trivial tree composed by one node (0-dimensional homology class of the image)

5   AT-RIG: Homological Tool or Topological Invariant?

The proof of homology and homotopy invariance of the AT-RIG is an issue out of the scope of this paper. The first part of this section is employed in supporting the thesis that the AT-RIG notion allows us to discriminate two non homologically equivalent objects having the same Betti numbers. Different instances (configurations spheres with handles, Menger sponges, torus, double torus, etc) are successfully examined with specific AT-segmentations. We only show here the example of AT-segmentations of simple cellular versions of the torus and the sphere with two handles.

On the other hand, we have only implemented software that calculates AT-segmentations (based on AT-models) but not AT-RIGs. Due to this reason, we expand the second part of this section to evaluate AT-segmentations of digital objects with known homology.

Given a torus (see Fig. 4), Fig. 5(a) shows a primal AT-segmentation, Fig. 5(b) an associated dual AT-segmentation, Fig. 6(a) its primal AT-RIG and Fig. 6(b) its dual AT-RIG.

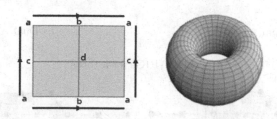

Fig. 4. Torus identification space

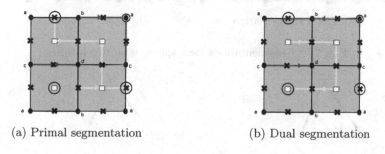

(a) Primal segmentation   (b) Dual segmentation

Fig. 5. Torus primal and dual segmentations

In Fig. 7, given a cell model of a sphere with two handles, we provide a primal AT-segmentation Fig. 7(a), an associated dual AT-segmentation Fig. 7(b), its primal AT-RIG Fig. 8(a) and its dual AT-RIG Fig. 8(b).

Figure 9 shows an example of partition of a primal AT-segmentation of the Menger Sponge of recursion depth 2 [29], computed from an AT-model. The left side shows a Menger sponge with 400 0-cells, 1224 1-cells, 1056 2-cells and 312 3-cells. On the right side, a primal AT-partition is shown. The segmentation region of dimension $(0, 1)$ is shown in red. The regions of dimension $(1, 2)$ and $(2, 3)$ are shown in yellow and blue respectively. The second order primal AT-RIG is a star-type tree having as center the 0-dimensional homology generator (red region) and as leafs the 1-dimensional 81 homology classes (yellow regions R with $crt(R) > 0$). The blue regions do not appear in the second order primal AT-RIG due to the fact that their criticality number is zero. A similar example is shown in the lower part of Fig. 9 with a double torus. This example is composed by 714 0-cells, 1728 1-cells, 1280 2-cells and 268 3-cells. The second order primal AT-RIG

(a) Primal AT-RIG   (b) Dual AT-RIG

Fig. 6. AT-RIGs for a torus

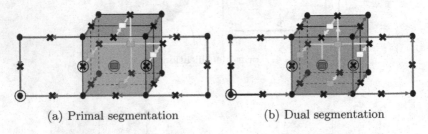

(a) Primal segmentation   (b) Dual segmentation

Fig. 7. AT-segmentations for a sphere with two handles

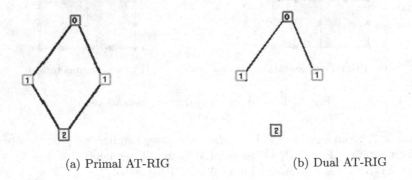

(a) Primal AT-RIG   (b) Dual AT-RIG

Fig. 8. AT-RIGs for a sphere with two handles

for this example has a 0-dimensional homology generator (red region) connected to 4 1-dimensional homology classes (yellow regions R with $crt(R) > 0$) and all of them connected to a single 2-dimensional homology class representing the cavities (blue region R with $crt(R) > 0$).

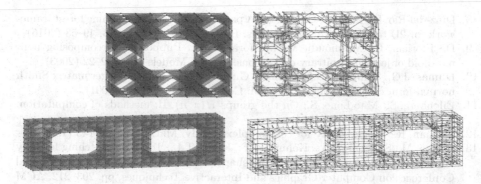

Fig. 9. (Left) A Menger Sponge of recursion depth 2 and a double torus (Right) Result of their respective primal AT-partitions, where the region of dimension $(0,1)$ is colored in red, $(1,2)$-regions in yellow and $(2,3)$ regions in blue. (Color figure online)

6   Conclusions

In this paper, a new topological tool, called second order AT-RIG, for distinguishing cell complexes beyond Betti numbers or Euler characteristic is algorithmically designed. This tool allows to discover relationships between homology classes of dimension differing in one. The concise experimentation carried out supports the hypothesis that the AT-RIG is a well-defined notion and that these relations are "up to homology". To theoretically prove these results would suppose a true revolution in the field of the topological representation. Anyway, negative answers would still mean a step forward because we have a useful topological tool properly working within an AT-model context.

References

1. Alexandroff, P.S.: Combinatorial Topology. Dover, New York (1998)
2. Ayala, R., Domínguez, E., Francés, A.R., Quintero, A.: Homotopy in digital spaces. In: Borgefors, G., Nyström, I., di Baja, G.S. (eds.) DGCI 2000. LNCS, vol. 1953, pp. 3–14. Springer, Heidelberg (2000). https://doi.org/10.1007/3-540-44438-6_1
3. Boykov, Y.Y., Jolly, M.P.: Interactive graph cuts for optimal boundary and region segmentation of objects in ND images. In: Proceedings of Eighth IEEE International Conference on Computer Vision, vol. 1, pp. 105–112 (2001)
4. Cadek, M., Krcal, M., Matousek, J., Vokrinek, L., Wagner, U.: Polynomial-time computation of homotopy groups and Postnikov systems in fixed dimension. SIAM J. Comput. **43**(5), 1728–1780 (2014)
5. Carr, H.A., Weber, G.H., Sewell, C.M., Ahrens, J.P.: Parallel peak pruning for scalable SMP contour tree computation. In: IEEE 6th Symposium on Large Data Analysis and Visualization (LDAV), pp. 75–84 (2016)
6. Couprie, M., Bertrand, G.: Asymmetric parallel 3D thinning scheme and algorithms based on isthmuses. Pattern Recogn. Lett. **76**, 22–31 (2016)
7. Delfinado, C.J.A., Edelsbrunner, H.: An incremental algorithm for Betti numbers of simplicial complexes on the 3-sphere. Comput. Aided Geom. Des. **12**(7), 771–784 (1995)

8. Díaz-del-Río, F., Real, P., Onchis, D.: A parallel homological spanning forest framework for 2D topological image analysis. Pattern Recogn. Lett. **83**, 49–58 (2016)
9. De Floriani, L., Mesmoudi, M.M., Morando, F., Puppo, E.: Decomposing nonmanifold objects in arbitrary dimensions. Graph. Models **65**(1), 2–22 (2003)
10. Dumas, J.G., Saunders, B.D., Villard, G.: On efficient sparse integer matrix Smith normal form computations. J. Symbol. Comput. **32**(1), 71–99 (2001)
11. Eilenberg, S., Mac Lane, S.: On the groups $H(\pi, n)$, II: methods of computation. Ann. Math. **60**, 49–139 (1954)
12. Forman, R.: Morse theory for cell complexes. Adv. Math. **134**, 90–145 (1998)
13. Hilaga, M., Shinagawa, Y., Kohmura, T., Kunii, T.L.: Topology matching for fully automatic similarity estimation of 3D shapes. In: Proceedings of the 28th Annual Conference on Computer Graphics and Interactive Techniques, pp. 203–212. ACM (2001)
14. De Floriani, L., Fugacci, U., Iuricich, F.: Homological shape analysis through discrete morse theory. In: Breuß, M., Bruckstein, A., Maragos, P., Wuhrer, S. (eds.) Perspectives in Shape Analysis. MV, pp. 187–209. Springer, Cham (2016). https://doi.org/10.1007/978-3-319-24726-7_9
15. Dumas, J.G., Heckenbach, F., Saunders, D., Welker, V.: Computing simplicial homology based on efficient Smith normal form algorithms. In: Joswig, M., Takayama, N. (eds.) Algebra, Geometry and Software Systems, pp. 177–206. Springer, Heidelberg (2003). https://doi.org/10.1007/978-3-662-05148-1_10
16. Fiorio, C.: A topologically consistent representation for image analysis: the frontiers topological graph. In: Miguet, S., Montanvert, A., Ubéda, S. (eds.) DGCI 1996. LNCS, vol. 1176, pp. 151–162. Springer, Heidelberg (1996). https://doi.org/10.1007/3-540-62005-2_13
17. González-Díaz, R., Real, P.: On the cohomology of 3D digital images. Discret. Appl. Math. **147**(2), 245–263 (2005)
18. González-Díaz, R., Jiménez, M.J., Medrano, B., Real, P.: Chain homotopies for object topological representations. Discret. Appl. Math. **157**(3), 490–499 (2009)
19. Gonzalez-Lorenzo, A., Bac, A., Mari, J.L., Real, P.: Allowing cycles in discrete Morse theory. Topol. Appl. **228**, 1–35 (2017)
20. Günther, D., Reininghaus, J., Wagner, H., Hotz, I.: Efficient computation of 3D Morse-Smale complexes and persistent homology using discrete Morse theory. Vis. Comput. **28**(10), 959–969 (2012)
21. Haarmann, J., Murphy, M.P., Peters, C.S., Staecker, P.C.: Homotopy equivalence in finite digital images. J. Math. Imaging Vis. **53**(3), 288–302 (2015)
22. Harker, S., Mischaikow, K., Mrozek, M., Nanda, V.: Discrete Morse theoretic algorithms for computing homology of complexes and maps. Found. Comput. Math. **14**(1), 151–184 (2014)
23. Hurewicz, W.: Homology and homotopy theory. In: Proceedings of the International Mathematical Congress, p. 344 (1950)
24. Klette, R.: Cell complexes through time. In: International Symposium on Optical Science and Technology, pp. 134–145. International Society for Optics and Photonics (2000)
25. Kong, T.Y., Rosenfeld, A.: Topological Algorithms for Digital Image Processing, vol. 19. Elsevier, Amsterdam (1996)
26. Kovalevsky, V.: Algorithms in digital geometry based on cellular topology. In: Klette, R., Žunić, J. (eds.) IWCIA 2004. LNCS, vol. 3322, pp. 366–393. Springer, Heidelberg (2004). https://doi.org/10.1007/978-3-540-30503-3_27
27. Lefschetz, S.: Algebraic Topology, American Mathematical Society Colloquium Publications, vol. 27. American Mathematical Society, New York (1942)

28. Lienhardt, P.: Topological models for boundary representation: a comparison with n-dimensional generalized maps. Comput. Aided Des. **23**(1), 59–82 (1991)
29. Menger, K.: Allgemeine Räume und Cartesische Räume, Teil I, Amsterdam, pp. 476–482 (1926)
30. Molina-Abril, H., Real, P., Nakamura, A., Klette, R.: Connectivity calculus of fractal polyhedrons. Pattern Recogn. **48**(4), 1150–1160 (2015)
31. Molina-Abril, H., Real, P.: Homological spanning forest framework for 2D image analysis. Ann. Math. Artif. Intell. **64**, 1–25 (2012)
32. Molina-Abril, H., Real, P.: Homological optimality in Discrete Morse Theory through chain homotopies. Pattern Recogn. Lett. **11**, 1501–1506 (2012)
33. Munkres, J.R.: Elements of Algebraic Topology. Addison-Wesley, Boston (1984)
34. Palmieri, J.H: Sage Module: Algebraic-Topological Model for a Cell Complex (2015). http://doc.sagemath.org/
35. Pilarczyk, P., Real, P.: Computation of cubical homology, cohomology and (co)homological operations via chain contractions. Adv. Comput. Math. **41**(1), 253–275 (2015)
36. Pudney, C.: Distance-ordered homotopic thinning: a skeletonization algorithm for 3D digital images. Comput. Vis. Image Underst. **72**(3), 404–413 (1998)
37. Real, P., Molina-Abril, H., Gonzalez-Lorenzo, A., Bac, A., Mari, J.L.: Searching combinatorial optimality using graph-based homology information. Appl. Algebra Eng. Commun. Comput. **26**(1–2), 103–120 (2015)
38. Real, P., Diaz-del-Rio, F., Onchis, D.: Toward parallel computation of dense homotopy skeletons for nD digital objects. In: Brimkov, V.E., Barneva, R.P. (eds.) IWCIA 2017. LNCS, vol. 10256, pp. 142–155. Springer, Cham (2017). https://doi.org/10.1007/978-3-319-59108-7_12
39. Romero, A., Rubio, J., Sergeraert, F.: Effective homology of filtered digital images. Pattern Recogn. Lett. **83**, 23–31 (2016)
40. Robins, V., Wood, P.J., Sheppard, A.P.: Theory and algorithms for constructing discrete Morse complexes from grayscale digital images. IEEE Trans. Pattern Anal. Mach. Intell. **33**(8), 1646–1658 (2011)
41. Saha, P.K., Borgefors, G., di Baja, G.S.: A survey on skeletonization algorithms and their applications. Pattern Recogn. Lett. **76**, 3–12 (2016)

Computing the Component-Labeling and the Adjacency Tree of a Binary Digital Image in Near Logarithmic-Time

Fernando Díaz del Río[1], Helena Molina-Abril[2(✉)], and Pedro Real[2]

[1] Department of Computer Architecture and Technology, University of Seville, Seville, Spain
[2] Department of Applied Mathematics, University of Seville, Seville, Spain
habril@us.es

Abstract. Connected component labeling (CCL) of binary images is one of the fundamental operations in real time applications. The adjacency tree (AdjT) of the connected components offers a region-based representation where each node represents a region which is surrounded by another region of the opposite color. In this paper, a fully parallel algorithm for computing the CCL and AdjT of a binary digital image is described and implemented, without the need of using any geometric information. The time complexity order for an image of $m \times n$ pixels under the assumption that a processing element exists for each pixel is near $O(log(m + n))$. Results for a multicore processor show a very good scalability until the so-called memory bandwidth bottleneck is reached. The inherent parallelism of our approach points to the direction that even better results will be obtained in other less classical computing architectures.

Keywords: Component-Labeling · Adjacency tree · Digital image Parallelism

1 Introduction

Connected component labeling (CCL) of binary images is one of the fundamental operations in real time applications, like fiducial recognition [6] or classifying objects as connected components (CCs). The labeling operation transforms a binary image into a symbolic matrix in which every element (pixel) belonging to a connected component is assigned to a unique label. Currently, there are mainly four classes of CCL algorithms: Multi-scan algorithms, Two-scan algorithms, Tracing-type algorithms and Hybrid algorithms mixing the previous ones. All of them (including the fastest one) use raster or Tracing-type approaches,

This work has been supported by the Spanish research projects MTM2016-81030-P (AEI/FEDER, UE) and TEC2012-37868-C04-02 of Ministerio de Economía y Competividad and the VPPI of the University of Seville.

© Springer Nature Switzerland AG 2019
R. Marfil et al. (Eds.): CTIC 2019, LNCS 11382, pp. 82–95, 2019.
https://doi.org/10.1007/978-3-030-10828-1_7

scanning the whole binary image or its contours in a sequential manner. They begin labeling the first pixel and so the second one as a function of the first pixel label. This local processing proceeds progressively until the last pixel is reached. This fact necessarily implies data dependencies between the labeling of one pixel and the previous one, which prevents these methods from using a pure parallel approach. In terms of time complexity, this means that linear order $O(N)$ (being N the number of pixels) cannot decrease independently of the number of available processing units.

In relation to the representation of digital objects or, alternatively, binary digital images, various topological models have been exhaustively used. Adjacency trees (also called topological, inclusion or homotopy trees [2,16,17], and here AdjT, for short) offer a classical region-based representation in terms of rooted tree of certain topological and spatial properties of the connected components in a binary image. Within an AdjT, each node represents a distinct foreground (FG) or background (BG) component, and an edge between two nodes means that one of them is surrounded by the other. The root in an AdjT always represents the unique BG component "surrounding" the image (if it does not exist, it can be artificially created) and two 2D binary digital images are topologically equivalent if and only if their AdjTs are equivalent. An example of an AdjT of the binary image in Fig. 2 is shown in Fig. 4 (left). Aside from image understanding [18] and mathematical morphology applications [7,10,15], AdjTs have encountered exploitation niches in geoinformatics, dermatoscopics image, biometrics, etc. (see [3,5,6] for instance). Therefore, finding fast algorithms for segmenting and computing the AdjT of a 2D digital binary image is crucial for solving important problems related to topological interrogation in the current technological context.

It is evident that the compression of those nodes of a CCL tree (CCLT) satisfying the neighboring condition "having the same color", directly yields to the AdjT. In this paper we present a novel method for computing CCL and AdjT but reducing the number of operations, so that computation time and memory consumption are sensibly decreased, whereas the degree of parallelism is extended to every single pixel.

2   Related Works

Parallel implementations for computing topological magnitudes can be achieved via two approaches. On the one hand, there is some space for parallelism when codifying scan or tracing-based CCL algorithms. These algorithms contain two main stages: the scanning phase where provisional labels are assigned to pixels depending on their neighbors, and some kind of union-find technique to collect label equivalence information in the previous assignment. For example, dividing the image into strips is a classical data partition technique for obtaining parallelism. The second stage must then use a more sophisticated union-find technique for the provisional labels to get to the CCL. There is also some room for parallelism in this phase, and many works have addressed different variations (see [8,10,15]) including tuning parallel algorithms for specific computers (see [1]).

On the other hand, topology is the ideal mathematical scenario for promoting parallelism in a natural way, although it drives to less classical parallelism approaches. The nature of the topological properties is essentially qualitative and local-to-global, having the additional advantage that its magnitudes are robust under deformations, translations and rotations. Nevertheless, the results in the literature in that sense are rare. Up to now, the only topological invariant that has been calculated using a fully parallel computation is the Euler number [3]. Other authors have recently proposed other parallel algorithms that compute some aspects of the homological properties of binary images [13]. In [4], a digital framework for parallel topological computation of 2D binary digital images based on a sub-pixel scenario was developed, modeling the image as a special abstract cell complex [11], in order to facilitate the generalization of this work to higher dimensional images. In addition, some software libraries of flexible C++ (RedHom [16]) have appeared for the efficient computation of the homology of sets. These libraries implement algorithms based on geometric and algebraic reduction methods.

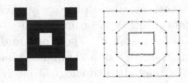

Fig. 1. Holes of 4-adjacent CCs are 8-adjacent CCs and vice versa for 2D binary digital images based on square pixel.

In relation to previous works, we construct our scaffolding on the basis of the two following basic topological properties: "being adjacent to" and "being surrounded by". Moreover, we take advantage of the powerful duality properties that the topological invariants of connected components and holes have in the context of 2D binary digital images based on square pixel. In other words, we exploit that the holes of 4-adjacent CCs must be 8-adjacent CCs and vice versa (see Fig. 1). As a result, all the algorithms of this work use simple connectivity graphs (CGs) as their basis. Our simplification allows us to reduce the number of operations, the computation time and the memory consumption, and to extend the degree of parallelism to every single pixel.

3   A Convenient Topological Framework for Computing CCL and AdjT

In a few words, topological analysis of digital images studies their degree of connectivity, defining fixed adjacency relations between pixels as "local neighborhood measures". It is obvious that a unique tree covering all the pixels of an image can be built (no matter the intensity of the pixels) going through all the pixels of the image using always, for example, the North direction, until the upper border is found and then change to East direction until the most northeastern pixel is reached. Pixels that connect different colors should be marked as candidates of frontiers between CCs (region frontiers). Any tree covering the image plus the region frontier candidates is then an instance of a connectivity tree that holds the complete information of the binary image. Note that for 2D binary digital images, there are two types of connectivity we must deal with, which are: CCs and holes. However, these two concepts can be reduced to one for a binary image since a hole can be seen as a CC that is surrounded by another CC of different color. Concerning topological analysis, all local neighboring conditions used here are derived from 4-adjacency relations. Concerning the output of this processing, the white nodes of the AdjT are 4-CCs and the black ones are 8-CCs. Let us limit ourselves to say that a hole of a 4-connected FG object can be interpreted as an 8-connected component of the BG. For this reason the algorithm presented here considers the FG with 4-adjacency and the pixels of the BG with 8-adjacency. Then, a minimal tree can be computed having always the correct number of FG 4-connected components and BG 8-connected components.

Using a combinatorial optimization process, it is possible to find a connectivity tree in which every CC has only one region frontier pixel. Each one of these special pixels marks a bond between two neighboring CCs of different color. For example, in Fig. 3 these pixels are marked with a number, which is the representative label of each CC. In the figure, black regions can be compressed to the pixels labeled with the numbers 92, 165, 194, 200, 226, 258, whereas white areas are represented by the pixels 49, 77, 179. We can say that these region frontier pixels are the "attractors" of each one of the trees that contain a CC. In this paper, a connectivity tree that holds this property is called a connected component labeling tree (CCLT). Following the path given by this compression (that is going along the connectivity tree) the different frontier pixels can be found. In the light of the above, in this paper a fully parallel algorithm for computing the CCL (and the AdjT) of a binary digital image based on square pixels is implemented through the building of one CCLT. This is achieved without the need of using any geometric information. Once the combinatorial optimization process has been carried out, the whole image (and, thus, every CC) can be compressed to just one pixel. Hence, the compression of those nodes of a CCLT whose neighboring condition is "having the same color" directly yields to the AdjT. To carry out this process, the local neighboring connectivity information (that is, those of the adjacent pixels) is first transformed into a global connectivity graph (CG) tracking a unique direction in a number of iterations equal to $log(m + n)$ (being m the image width, and n the image height). Finally, CCs are extracted by fusing the regions of the previous global graph in a parallel way using a reduced number of iterations (see Sect. 4). A more detailed example is shown in Fig. 2. On the left a face-like binary image is depicted, whereas on the right, its corresponding black (FG) 4-connected components are identified.

Note that 4-adjacency prevents the nose from being connected with the glasses. Each FG CC has been assigned a black representative pixel (marked with little triangles in Fig. 2 right), which can be called "attractor" as every link of the simplified CCLT collapses in them. Conversely, FG holes can be represented by attractors of the BG CCs. These are marked with little downwards arrows, and can be called BG attractors, for analogous reasons (that is, all the BG CC collapses in it). An additional BG attractor in the most left-bottom corner has been added to represent a virtual attractor for the whole image. Whereas the meaning of an attractor is intuitive and sufficient for understanding these concepts, an exact definition of the attractors is given in Sect. 4.

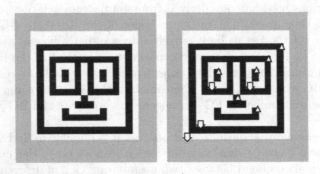

Fig. 2. Left: original face-like image. Right: FG attractors (little triangles) and BG attractors (downwards arrows).

The results for Fig. 3 (left) returned by our algorithm that generates the CCL and the AdjT (see next section) are summarized in the table in Fig. 3 (right) and Table 1. These tables contain the FG and BG ordered pairs. The silhouette contained in Fig. 2 has a total of 6 FG attractors and 4 BG attractors. For each FG attractor, FG attractor pair table (Fig. 3) gives the BG attractor that goes to (that is, the hole that surrounds this component) and some additional information (area and perimeter of each region). Numbers for attractors are pixel labels, that is, linear indexes over the image (following a column convention, that is, the linear indexes for a matrix when all its elements are numbered following its columns). The perimeter is computed by counting the 8-adjacent pixels that enclose a CC (which means that a CC composed of a single pixel has a perimeter of 8 pixels).

Likewise, BG attractor pair table (Table 1, left) gives the FG attractor where each BG attractor points to (that is, the black component that surrounds a hole). Note that AdjT can be automatically extracted from these tables. The matrix of Table 1 (right) is a possible compact representation of the AdjT and it gives the crossed including relations of FG and BG attractors using this notation: rows are FG attractor indexes (labels); columns are BG attractor labels; 1 means that a BG attractor is included on an FG attractor, −1 means that an FG attractor is included on a BG attractor. Finally Fig. 4 (left) comprises a graphical

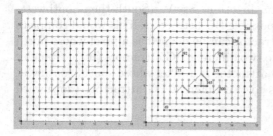

FG Attr. index	BG Attr. Index where it points to	FG CC area (in pixels)	FG CC perimeter (in pixels)
92	77	2	10
165	49	4	14
194	179	2	10
200	49	9	24
226	49	37	48
258	17	54	62

Fig. 3. Left: final CCLT having the minimal number of FG 4-connected components and BG 8-connected components. Indexes of BG/FG attractors are on the right of each attractor. Right: F_{TABLE}: the corresponding FG attractor table. Numbers are the linear indexes for a matrix numbered following its columns.

Table 1. Left: AdjT of Fig. 3. Rows are FG attractor labels; columns are BG attractor labels; values are: 1 = BG attractor included on the FG attractor, −1 = FG attractor included on the BG attractor. Right: B_{TABLE}: BG attractor table of Fig. 3. Numbers are the linear indexes for a matrix numbered following its columns

BG Attractor index	FG Atractor index where it points to	BG CC area (in pixels)	BG CC perimeter (in pixels)
49	258	82	54
77	226	10	18
179	226	10	18

	49	77	179
92	0	-1	0
165	-1	0	0
194	0	0	-1
200	-1	0	0
226	-1	1	1
258	1	0	0

representation of a weighted AdjT of Fig. 3. Filled (empty, resp.) circles are FG (BG, resp.) CC. The notation for the numbers $i : \{a, p\}$ represents: $i = index(label)$, $a = area$ of the CC, $p = perimeter$ of the CC.

4   A Parallel Algorithm for Building the CCLT

As previously explained, CG should pair any FG pixel with another FG pixel, except those that are the possible attractors of a CC. The key point is that each pairing must be done in a convenient direction so that only one unpaired pixel (the attractor) exists for each CC. In this case, if we followed the links from any pixel, this stream of links would fall on this attractor. In Fig. 5 (left) a simple shape is depicted where every FG pixel has been linked following a simple criterion: if its North neighbor were an FG pixel, it would be linked to it; if it were not, it would be linked with its East adjacent pixel if it had the FG color; if East adjacent pixel had not FG color either, it is marked as a possible attractor and connected to the north BG pixel. This pairing is called here "North-then-East criterion" or simply NE criterion. For the 8-adjacent BG pixels it is convenient that the criterion uses the opposite direction: SswW (South-then-Southwest-then-West), thus obtaining a set of possible BG attractors. This would complete a possible CG. Hence, a false FG attractor can be defined as a pixel whose north and east adjacent BG pixels points to two different BG attractors (little triangles in Fig. 7, left). Likewise, a false BG attractor is a pixel whose south, southwest and west adjacent FG pixels points to two different FG attractors (downward arrows in Fig. 7, left).

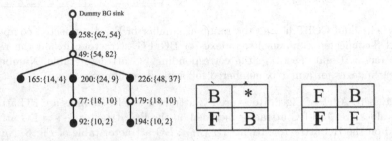

Fig. 4. Left: a graphical representation of the AdjT of Fig. 3. Filled (empty, resp.) circles are FG (BG, resp.) CCs. Notation $i : \{a, p\}$ represents: i = index (label), a = area of the CC, p = perimeter of the CC. Right: The 2×2 patterns that represent attractors in F*. B* matrices are resp. (pixel of reference in bold).

However, things are not so simple in the general case, because many pixels can have both: a North and an East neighbor, and only one must be selected for the pairing. For instance, the right picture of Fig. 5 shows a spiral shape where the direction of the every pairing was done in an ad-hoc form, so that only one unpaired pixel remains (the most northeast FG pixel). Note that the NE criterion is not preserved for many FG pixels (correspondingly with respect to the SswW criterion for BG pixels). The same for the BG: its only unpaired pixel is the dummy attractor on the most southwest corner. The key is how these directions are selected in parallel to produce the desired pairing. Global information about the shape of every CC is needed to choose this correctly, that is, it is impossible a priori to discover which pairing must select every pixel to get to the correct CCLT. Nevertheless, there exists a high amount of parallelism in this process. In order to get to the CCLT, we propose two main steps: Generating a CG as parallel as possible; and secondly, transforming this CG into a correct CCLT through the cancellation of pairs of a false FG attractor with a BG one. This process must be iterated until no false attractor remains.

As stated before, an algorithm that tries to extract global information of an image must include some pieces to search the relation between remote parts of the image. Using the properties of the tree-like structures, those sequential pieces can be reduced significantly. Figure 6 draws a sketch of the process to obtain the minimal tree structures needed to extract the CCLT and AdjT of an image, preserving its combinatorial nature. From the image I, a local CG based on the local information of each pixel can be first computed. Here "local" means that

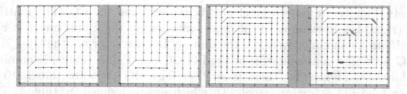

Fig. 5. Two figures with one CC. Left: a simple shape where the pixel pairing is accurate by NE criterion. Right: a spiral, where this simple criterion is not valid.

the computation of every link is based only on the values of its (e.g. 4-) adjacent pixels. This tree will contain the links from each pixel to its immediate neighbors. Then, through successive iterations we can get to a different global CG. Here "global" means that each pixel knows the link to its (possibly far) attractor in this CG. FG, BG attractors of the initial CG are not yet the true attractors (see Fig. 7, where the FG CC has resulted in two attractors; one of them must be false). This tree must be transformed so as to contain only true attractors, which means that we have reached the correct CCLT. Using the CCLT, every attractor can be related with another attractor of opposite color that contains the first. This is a representation of the AdjT.

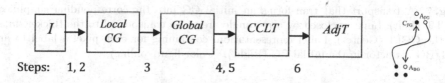

Fig. 6. Left: steps involved in a sequential CCLT building. Right: a cycle from an FG attractor A_{FG} to C_{FG} through a BG attractor A_{BG}, where $C_{FG} = A_{FG}$.

From now on, let us suppose that the border of the whole image is composed of BG pixels, which belongs to an external dummy BG attractor. The aim is to build an optimal gradient vector field with only one (FG or BG) attractor for each (resp. FG or BG) CC. The first step of Fig. 6 computes an initial CG of the image I. The computation of every link is exclusively based on the values of its adjacent pixels. The rest of the steps are necessary to transform this CG into a CCLT detecting the representative FG, BG attractors. The second step determines which pixels are possible BG/FG attractors, that is, those that have a link that connects FG and BG pixels. The key point is that the FG graph must be built on the opposite direction than that of the BG. Next, the third step introduces global relations between pixels and attractors, so that the attractor for each pixel is determined when following the vector field of the CG. In sequential form, a pixel can track its links and then check if its neighbor pixel is an attractor. If not, this operation would be repeated for the next neighbor and so on, until an attractor is reached. Each pixel can store a label of the attractor to which it points. Finally, we have a label matrix representing the CG. The matrix of Fig. 7 (right) is ann example of this representation for the simple image in Fig. 7 (left). Using a column convention, the dummy BG attractor has the label 7, and the FG attractor, representative of the FG CC (see Fig. 7, center) is numbered with label 38. Meanwhile, there is another false BG attractor (label 18) and another false FG attractor (label 23), which are underlined in Table 4. These false attractors must be coupled (step 4 of Fig. 6) for the final CCLT, so that the underlined label 23 would be substituted by 38, and the underlined label 18 for 7. The next step consists of transports, or equivalently the fusion of those parts of a same CC, performed by a CG combinatorial optimization process in order to get a tree that has as many nodes as 4-connected components the image has.

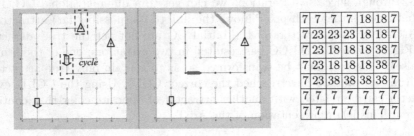

Fig. 7. A transport that transforms an initial CG into the corresponding simplified CCLT (Left). Links enclosed by a rectangle are to be transported to the thicker links in the CCLT (Center). Label representation, containing for each pixel a label to an FG/BG attractor in the initial CG (Right). (Color figure online)

Graphically, a simple conversion of an initial CG (left) into the CCLT (center) is shown in Fig. 7. Note that the CG of Fig. 7 (left) has one cycle (see the red and green edges surrounding the word "cycle"). Only one cancellation of a pair of false attractors is needed to get the CCLT. CG links that are enclosed by a dotted rectangle are transported to the thicker links for the CCLT. FG attractors are depicted with little triangles, whereas the BG attractors with downward arrows. It can be easily shown that selecting opposite directions for the BG pixel and for the FG pixels when building the CG implies that every BG attractor breaks an FG CC, and vice versa. Thus, by canceling FG-BG attractor pairs until only one attractor would remain for each FG and BG CC, so the transformation from CG into the CCLT is accomplished. After this process there must be only one attractor for each FG (resp. BG) CC. It is worth to note that the process of link transporting is done exclusively handling the CG. The links in the CG enclosed by a dotted rectangle in Fig. 7 (left) are transported, in such a way that both false FG and BG attractors disappear. This is depicted with thicker links in the CCLT (Fig. 7, center). Any transport implies the re-labeling of the label representation (like the matrix of Fig. 7, right). Note that finally the remaining BG attractor is located on the SW corner of the image. The couples to be cancelled can be found by following a path along the CG and by transporting its corresponding links. Yet more, it can be shown that most of these cancellations can be done in parallel, as demonstrated below.

Finally these attractors will define the AdjT in a straightforward form without any geometric computation: Simply, each FG attractor is connected through the CCLT to another BG attractor. And vice versa: each BG attractor is linked to another FG attractor (Fig. 3). So the question is now: what parts of a CCLT building can be done in parallel for the huge amount of pixels that a digital image can have? Whereas first two steps of Fig. 6 are independent for every pixel (thus trivially parallel), the crucial step Fig. 6 requires in principle a sequential processing. Nevertheless, most of the attractors can be coupled in parallel if next properties are taken into account. Let us consider the FG attractor A_{FG} in Fig. 6 (right). The adjacent East BG pixel of A_{FG} fell (going to South direction along the BG path defined by the CG) to a BG attractor B_{FG}. Likewise, the adjacent West FG pixel of B_{FG} arrives (going to North direction along the FG path of the CG) to an FG attractor CFG. All the pairs of FB and BG attractors that fulfill A_{FG} = CFG can be cancelled in parallel, because (a) there are BG and FG paths that connect them, and (b) any tree structure has a unique root. Due to (a) the link of the false FG attractor can be transported so as to join the two FG pixels that the BG attractor was separating. Likewise for the link of the false BG attractor. These are the transports from Fig. 7, left to center. Moreover because of (b), there cannot exist two false FG attractors that use the same false BG attractor to be cancelled.

The parallel Algorithm 1 consists of the following steps. From image I, the possible attractors based on the local information of each pixel can be fully determined in parallel, and the same for the initial CG, for example using the North-then-East criterion (steps 1–4 of Algorithm 1). Using this local information and through successive iterations, the global CG can be obtained, which corresponds to the steps 5 to 9 of Algorithm 1). Now the possible FG, BG attractors can be efficiently coupled in parallel (steps 10–21 of Algorithm 1). At the end of this stage, we obtain the CCLT comprised in the final pointer matrix P of Algorithm 1, and in the true attractors, one for each FG and BG CC. Note that BG attractors now are the holes of the FG CCs (and vice versa). Finally, by means of steps 22–23, inclusion relations between BG and FG CCs can be extracted from the label pointed by each attractor.

As our aim is to describe the inherent parallelism that can be exploited in the CCL tree building, the notation followed here describes the algorithm in an OCTAVE/MATLAB-like form, which indicates in an direct way what are the data parallelism and the real data dependences. Therefore it is evident how each sentence could be implemented in a SIMD processor (or in SIMD kernels) or in SIMT oriented GPUs. Also OpenMP codes can be written almost directly through this notation, just by transforming each matrix operation into two nested loops, the outer of which can be commanded by a directive #pragma omp parallel for. An additional advantage is that the memory access patterns can be clearly observed with this notation. This can give a fast idea of the computing times because memory access is currently the most important bottleneck in current multicore processors [14,21]. For similar reasons, those sentences that can be executed in parallel (which have no real data dependences) are grouped in the same step. For example, the range of elements that can be processed in parallel is shown for each matrix (vector) with the notation $A(1:m,1:n)$. This means that the operation is extended all over the elements in rows $1,2,3,\ldots,m$ and in columns $1,2,3,\ldots,n$. Furthermore, we have avoided those matrix operations that cannot be done in an element-by-element way (like matrix inversions, matrix multiplication, etc.). Nevertheless, matrix operations that can be executed in a fully parallel form are introduced with the OCTAVE/MATLAB notation (e.g. $A.*B$ means an element-by-element multiply). Therefore, only one loop "for" and another one loop "while" that present dependences among its iterations are encountered. In addition, the algorithm does not have any conditional sentence. Some auxiliary matrices and predicative-like code, have transformed conditional operations into element-by-element logical ANDs or multiply operations for the possible results. This also promotes efficiency when using SIMD kernel codification and prevents the so-called thread divergence for GPUs, promoting a better performance on these platforms [14]. Supposing an image of m × n pixels and p processing elements, time complexity order can be trivially obtained according to the notation of Algorithm 1, and because most of the operations are done in an element-by-element form. Steps 1–4 are of this kind, so their time complexity is $O(m \times n/p)$. Steps 6–8 proceed in the same manner, but they are surrounded by a "for" loop (steps 5 and 9) with $log2(m+n-1)$ iterations. Thus, their complexity is $O((m \times n/p) \times log(m+n))$. Next step 10 can be computed fully in parallel, which supposes a complexity of $O(m \times n/p)$. Step 15 is similar to 10, but because many false attractors were previously deleted from matrixes $F*$ and $B*$ in 10, each "surviving" possible attractor has to be found. This supposes a searching of a variable length s10 that depends on the characteristic of the image. Steps 17 and 19 have the same complexity of 10 with different lengths s17 and s19, respectively. Moreover, steps 17 and 19 are enclosed by a while loop (16 and 21), with a number of iterations q, that is in general very little. These searching lengths s10, s17 and s19, and the number of iterations q can be related with image shapes; but for the random images of different densities [22], maximum values $smax = max\{s10,max,s17,max,s19,max,s22,max\}$ were very low: 72 for 512 × 512 pixels, 94 for 1 Mpixels and 176 for 4 Mpixels. Besides, q reached 3 only for one of the random images, whereas was 1 for all the tested real ones. In other words, the most time consuming steps are 5 to 9. Summarizing previous steps 1–21, it can be stated that, under the assumption that a processing element exists for each pixel $(p = m \times n)$, time complexity order for computing the CCL is very near to the logarithm of the width plus the height of the image. Step 22 involves only the true FG and BG CC (namely v), and can be done in parallel for any CC because matrix P comprises all the connectivity information. The only iterative procedure here is again the number of hops needed to find a true attractor. Thus, the time complexity order for this step is $O(s22 \times v/p + s22, max)$, where $s22, max$ is the maximum number of hops along the pointer matrix P to find an attractor. Step 23 can be done in parallel, being $O(v/p)$ its time complexity. To sum up, time complexity order for computing the CCL and the AdjT, under the assumption that a processing element exists for each pixel, is $O(log(m + n)) + O(q \times smax)$.

5   Testing Results and Conclusions

A complete implementation was done in C++/OpenMP through a direct translation of Algorithm 1. The compiler was Microsoft Visual Express. The server where tests were carried out was an Intel Xeon E5 2650 v2 with: 2.6 GHz, 8 cores, 8×32 KB data caches, Level 2 cache size 8×256 KB, Level 3 cache size 20 MB, maximum RAM bandwidth: 59.7 GB/s. Experiments were run 25 times and minimum times were collected, because this server runs concurrently lots of processes and this increases unfairly timing measurements. However, mean times differs only by a 10% wrt. to the minimum ones. Figure 8 shows the results for random images (taken from YACCLLAB [22]) with different sizes and densities (percentage of FG pixels). Although absolute computation times (being between 0.3 and 0.5 s for 512×512 images of different densities) are not faster than that of YACCLAB, this comparison is not fair since our method computes both black and white CCs, thus having a complete representation in terms of the AdjT, whereas classical CCL methods return only black CC labels. Nevertheless, speedup (time for various threads divided by time for 1 thread) is near the number of threads (Fig. 8), which points out that achieved scalability is very satisfactory for all image sizes and densities.

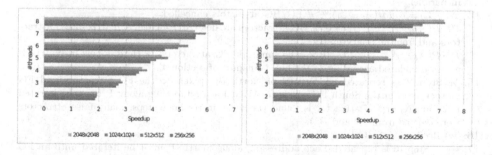

Fig. 8. Left: speedup for 1 to 8 threads for images of different sizes and: Right: density of 0.9. Left: density of 0.4.

In future works we will define more formally the notions of our algorithm so that additional properties will be exhibited. This would also serve to extend our method to bigger dimensions.

Algorithm 1. Given a binary matrix I, computes $P(CCL), B_{TABLE}, F_{TABLE}$ (column 1: index of the BG/FG attractor resp.; column 2: index of the FG/BG attractor, resp.), and AdjT. B means BG and F FG value

---

1: $I_{CC} \leftarrow I(2:m-1,2:n-1)$; % Central matrix
   $I_{NC} \leftarrow I(1:m-2,2:n-1)$; % North adjacent matrix. Similar for other 4-adjacent matrices I_{EC}, I_{SC}, I_{WC} (East, South, West) and 8-adjacent matrices $I_{SW}, I_{SE}, I_{NW}, I_{NE}$

2: $F^* \leftarrow (I_{NC} == BG).*(I_{EC} == BG).*(I_{CC} == FG)$;
   $B^* \leftarrow (I_{NC} == FG).*(I_{NE} == BG).*(I_{CC} == FG).*(I_{EC} == FG)$;
   % See 2x2 patterns in Figure 4 (right)
3: $R, C \leftarrow ndgrid(1:m, 1:n)$; % auxiliary matrices that contain a grid of row and column indexes.
4: $X_{NE} \leftarrow (I_{NC} == BG).*(I_{EC} == FG).*(I_{CC} == FG)$;
   $Y_{NE} \leftarrow (I_{NC} == FG).*(I_{CC} == FG)$ % initial X,Y directions
   $P \leftarrow F^*.*((C-1)*m + R)$; % initial local CG as an NE pointer matrix. Only attractors are set with column indexes.
5: **for** $k = 1 : log2(m+n-1)$ **do**
6:    $R_{hop,NE} \leftarrow R - Y_{NE}$;
      $C_{hop,NE} C + X_{NE}$; % row, column indexes are "moved" to the North or East according to X_{NE}, Y_{NE} values.
7:    $L_{hop,NE} \leftarrow sub2ind(Y_{NE}, R_{hop,NE}, C_{hop,NE})$; % R, C matrices are converted into column indexes.
8:    $P \leftarrow P(L_{hop,NE})$; % pointer matrix is updated.
      $X_{NE} \leftarrow X_{NE} + X_{NE}(L_{hop,NE})$;
      $Y_{NE} \leftarrow Y_{NE} + Y_{NE}(L_{hop,NE})$; % XNE, YNE are updated
9: **end for**% After this loop, P contains the global CG
10: %First coupling using East BG pixel to each FG attractor.
    $A_{FG} \leftarrow P(F^*(2:m-1,2:n-1))$; % Vector of FG attractors
    $A_{FG,East} \leftarrow P(F^*(2:m-1,3:n))$;% East BG pixels to FG attractors
11: $A_{BG} \leftarrow P(A_{FG,East})$; % Vector of BG attractors
12: $C_{FG} \leftarrow P(A_{BG} + m)$; %FG attractors from West FG ABG pixels
13: $A_{cancel} = (A_{FG} == C_{FG})$; % Logical Vector of FG/BG attractors that must be cancelled in parallel.
14: $N_{cancel} = count(A_{cancel})$;% # FG/BG attractors cancelled.
    $B^*(P(A_{BG}(A_{cancel}))) \leftarrow 0$; % BG attractors are deleted from logical matrix of BG attractors and from auxiliary matrices.
    $F^*(P(A_{FG}(A_{cancel}))) \leftarrow 0$; % The same for FG attractors. % Here, labels in matrix P are also updated according to the link transport of section 4.
15: % Step 10 is repeated for Vectors of North BG pixels to the FG attractors and FG attractors from the South FG pixels to ABG. Each addressing along P must be iterated until an FG or BG attractor is found (because, in previous steps, many false attractors were deleted from F^* and B^*).
16: **while** $N_{cancel} > 0$ **do**
17:    % Step 10 is repeated. Each addressing along matrix P must be iterated until an FG or BG attractor is found.
18:    $N_{cancel} = count(A_{cancel})$
19:    % Step 15 is repeated. Again addressing along matrix P must be iterated until an FG or BG attractor is found.
20:    $N_{cancel} = count(A_{cancel}) + N_{cancel}$; % total number of FG/BG attractors cancelled in current "while" iteration.
21: **end while**
22: % Extract attractor pair tables F_{TABLE}, B_{TABLE} from previous attractors using P.
23: % Compute AdjT using attractor pair tables F_{TABLE}, B_{TABLE}

References

1. Bhattacharya, P.: Connected component labeling for binary images on a reconfigurable mesh architecture. J. Syst. Arch. **42**(4), 309–313 (1996)
2. Buneman, O.P.: A grammar for the topological analysis of plane figures. Mach. Intell. **15**, 383–393 (1969)
3. Chiavetta, F., Di Gesù, V.: Parallel computation of the Euler number via connectivity graph. Pattern Recognit. Lett. **14**, 849–859 (1993)
4. Diaz-del-Rio, F., Real, P., Onchis, D.: A parallel homological spanning forest framework for 2D topological image analysis. Pattern Recognit. Lett. **83**, 49–58 (2016)
5. Cohn, A., Bennett, B., Gooday, J., Gotts, N.: Qualitative spacial representation and reasoning with the region connection calculus. GeoInformatica **1**(3), 275–316 (1997)
6. Costanza, E., Robinson, J.: A region adjacency tree approach to the detection and design of fiducials. Video Vis. Graph., 63–99 (2003)
7. Cucchiara, R., Grana, C., Prati, A., Seidenari, S., Pellacani, G.: Building the topological tree by recursive FCM color clustering. In: 16th IEEE ICPR, vol. 1, pp. 759–762 (2002)
8. Gupta, S., Palsetia, D., Patwary, M.M.A., Agrawal, A., Choudhary, A.N.: A new parallel algorithm for two-pass connected component labeling. In: IEEE IPDP Symposium, pp. 1355–1362 (2014)
9. Heijmans, H.J.: Connected morphological operators for binary images. Comput. Vis. Imag. Understand. **73**(1), 99–120 (1999)
10. Kalentev, O., Rai, A., Kemnitz, S., Schneider, R.: Connected component labeling on a 2D grid using CUDA. J. Parallel Distrib. Comput. **71**, 615–620 (2011)
11. Kovalevsky, V.: Algorithms in digital geometry based on cellular topology. In: Klette, R., Žunić, J. (eds.) IWCIA 2004. LNCS, vol. 3322, pp. 366–393. Springer, Heidelberg (2004). https://doi.org/10.1007/978-3-540-30503-3_27
12. Keshet, R.: Shape-tree semilattice. J. Math. Imag. Vis. **22**(2–3), 309–331 (2005)
13. Murty, A., Natarajan, V., Vadhiyar, S.: Efficient homology computations on multicore and manycore systems. In: 20th Annual International Conference on High Performance Computing, pp. 333–342 (2013)
14. Oxley, J.G.: Matroid Theory, vol. 3. Oxford University Press, Oxford (2017). NVIDIA, Cuda C best practices guide version. http://developer.nvidia.com/
15. Patwary, M., Ali, M., Refsnes, P., Manne, F.: Multi-core spanning forest algorithms using the disjoint-set data structure. In: 26th IEEE IPDP Symposium, pp. 827–835 (2012)
16. Institute of Computer Science, Jagiellonian University (2017). REDHOM, Redhom. http://redhom.ii.uj.edu.pl/
17. Ranwez, V., Soille, P.: Order independent homotopic thinning for binary and grey tone anchored skeletons. Pattern Recognit. Lett. **23**(6), 687–702 (2002)
18. Rosenfeld, A.: Adjacency in digital pictures. Inf. Control **26**, 24–33 (1974)
19. Serra, J.: Image Analysis and Mathematical Morphology. Academic Press, Cambridge (1982)
20. Stell, J., Worboys, M.: Relations between adjacency trees. Theor. Comput. Sci. **412**(34), 4452–4468 (2011)
21. Williams, S., Waterman, A., Patterson, D.A.: Roofline: an insightful visual performance model for multicore architectures. Commun. ACM **52**, 65–76 (2009)
22. YACCLAB - Yet Another Connected Components Labeling Benchmark (2017). https://github.com/prittt/YACCLAB

Towards Emotion Recognition:
A Persistent Entropy Application

Rocio Gonzalez-Diaz[1], Eduardo Paluzo-Hidalgo[1]([⊠]), and José F. Quesada[2]

[1] Department of Applied Mathematics I, University of Seville, Seville, Spain
[2] Department of Computer Science and Artificial Intelligence, University of Seville, Seville, Spain
{rogodi,epaluzo,jquesada}@us.es

Abstract. Emotion recognition and classification is a very active area of research. In this paper, we present a first approach to emotion classification using persistent entropy and support vector machines. A topology-based model is applied to obtain a single real number from each raw signal. These data are used as input of a support vector machine to classify signals into 8 different emotions (neutral, calm, happy, sad, angry, fearful, disgust and surprised).

Keywords: Persistent homology · Persistent entropy
Emotion recognition · Support vector machine

1 Introduction

Emotion recognition is not a trivial task and different approaches have been explored so far (see for example [12]). Additionally, its applications are really important, such as gathering and processing satisfaction feedback in customers' services, generating statistical studies over a population, using emotion recognition to improve spoken language understanding during a conversation. Furthermore, it can help in human interaction as in KRISTINA project[1], where emotion recognition is applied in order to help the interaction between health professionals and migrated patients. Among the different theories about emotions proposed in the specialized literature, we follow the model described in [10,14], where a discrete theory of emotions is given, differentiating several basic groups of emotions (neutral, happy, sad and surprised) and organizing them in a spatial model. In [16] a review of different emotional speech recognition techniques can be consulted.

Topological data analysis is a well substantiated field useful to extract information from data (see [17]). Concretely, a recent tool in this area called *persistent entropy* has been successfully applied to distinguish discrete piecewise-linear functions (see [13]).

---

[1] http://kristina-project.eu/en/.

© Springer Nature Switzerland AG 2019
R. Marfil et al. (Eds.): CTIC 2019, LNCS 11382, pp. 96–109, 2019.
https://doi.org/10.1007/978-3-030-10828-1_8

In this paper, persistent entropy is used to model arousal (i.e., emotional state) and emotion recognition as follows. First, speech signals are considered as piecewise linear functions. Second, persistent entropy is computed from the lower-star filtration obtained from these functions. This persistent entropy embedding can be considered as a summary of the features that appear in raw signals, as intensity and intonation. The stability theorem for persistent entropy computed from lower-star filtrations [13] guarantees right comparison between signals and robustness against noise. Finally, a support vector machine is used to classify emotions via persistent entropy values. As far as our knowledge, no topology approaches have been previously applied to emotion recognition.

This paper is organized as follows: Basic emotion theory, the notions of persistent homology and persistent entropy, and machine learning knowledge required for the model are introduced in Sect. 2. In Sect. 3, the methodology followed in the experiments is explained. Results obtained from different training approaches are shown in Sect. 4. Finally, Sect. 5 provides conclusions and future work ideas.

2 Background

In this paper, different tools are mixed up in order to propose a unified and coherent framework for emotion classification. In this section, the basic concepts about acoustics, topology, machine learning and statistics are introduced.

Acoustic and Psychoacoustic Features. Emotions constitute the main field largely studied by psychologists. Following [14], we consider that emotions can be modeled spatially in a circle, being arousal and valence their main characteristic features. Accordingly, prosodic attributes of speech [7] are strongly related with emotion recognition. This research area takes into account several features of speech, in conjunction with gesticulation of the speaker. Some of those features are: pitch signal, number of harmonics, vocal tract, and speech energy.

Along this paper, just the physical features of the acoustic signal along with the processing results available from this signals (such as the contour of speech signal which is a feature affected by the arousal of the speaker), will be taken into account. The inclusion of visual features will be proposed in Sect. 5 as a natural continuation of this research. Sentences will be processed, assuming that certain attributes, like the fundamental frequency, intensity and duration, of a sound are meaningful for emotion production and recognition. These attributes are encapsulated under the notion of prosody. Depending on the prosodic pattern, a sentence can have very different emotional features. For example, happiness is linked usually with large fundamental frequency and, loudness, in contrast with sadness, normally related to the opposite. For further explanations about psychoacoustics, [8] can be consulted.

In the literature, some emotion classification techniques have been proposed (see [18]). Some of them employ prosody contours information of speech in order to recognize emotions, as, for example: artificial neural networks, the multichannel hidden Markov model, and the mixture of hidden Markov models. For a further approximation to paralinguistic theory see [15].

Topology Background. Topological data analysis (TDA) studies the *shape of data*. In our case, we apply topological data analysis tools to distinguish between piecewise linear function shapes. For an introduction to topological data analysis, [5] can be consulted.

Persistent entropy is the main tool from TDA that will be used in this paper. It sums up persistent homology information which "measures" homological features of shapes and of functions.

Informally, homology provides the number of n-dimensional holes, called the n-th Betti numbers and denoted by β_n. Intuitively, β_0 is the number of connected components, β_1 the number of tunnels and β_2 the number of cavities. However, for dimensions higher than 2, we lose the intuition about what a hole is.

Definition 1 (Betti number, informal, [2]). *If X is a topological space, then $H_n(X) \simeq \mathbb{Z}^{\beta_n}$ is called the n-th homology group of X if the power β_n is the number of independent n-dimensional 'holes' in X. We call β_n the n-th Betti number of X. Finally, the homology of X is defined as $H(X) = \{H_n(X)\}_{n=0}^{\infty}$.*

Observe that the concept of homology is not useful in practice. For example, suppose a dataset V of 10 points sampling a circumference. We expect that $H_0(V) \simeq \mathbb{Z}$ since a circumference has one connected component. However, the exact 0-th homology of V is $\mathbb{Z}^{10}$. Therefore, we need a tool to compute the homology of the underlying space sampled by a dataset. Following this idea, Edelsbrunner et al. [5] introduced the concept of persistent homology together with an efficient algorithm and its visualization as a persistence diagram. Carlsson et al. [20] reformulated and extended the initial definition and gave an equivalent visualization method called persistence barcodes.

Given a dataset V and a simplicial complex K constructed from it, persistent homology measures homology by a filtration during time, obtaining births and deaths of each homology class ('hole'). Consequently, those classes that persist are better candidates to be representatives of the homology of the underlying space.

Definition 2 (Abstract simplicial complex). *Let V be a finite set. A family K of subsets of V is an abstract simplicial complex if for every subsets $\sigma \in K$ and $\mu \subset V$, we have that $\mu \subset \sigma$ implies $\mu \in K$. A subset in K of $m+1$ elements of V is called a m-simplex and V is called the set of vertices of K.*

Definition 3 (Filtration). *Given a set V and a simplicial complex K constructed from it, a filtration is a finite increasing sequence of simplicial complexes:*

$$\emptyset = K_0 \subset K_1 \subset K_2 \subset \cdots \subset K_n = K$$

A particular filtration that will be used in this paper is the lower-star filtration.

Definition 4 (Lower-star filtration [5]). *Let K be a simplicial complex with real (distinct) values specified on the set V of all the vertices in K. Since vertices have distinct function values, then they can be ordered incrementally:*

$$f(u_1) < f(u_2) < \cdots < f(u_n).$$

The lower star of u_i is the subset of simplices of K for which u_i is the vertex with maximum function value,

$$K_i = \{\sigma \in K : \text{for all vertex } v \text{ of } \sigma \Rightarrow f(v) \leq f(u_i)\}.$$

Once the lower-star filtration is obtained, persistent homology can be computed as follows. The inclusion $K_i \subset K_j$ induces a homomorphism $f^{i,j}$: $H(K_i) \rightarrow H(K_j)$ on homology. Its image is the persistent homology, letting β_p^{ij} be the number of n-dimensional 'holes' that are born at K_i and die entering K_j. During the computation of persistent homology along the filtration, an elder rule is applied. For example, when there are two connected components that get joined at some K_j, the older one (the one that was born earlier) remains, and the younger one dies. A persistence barcode is a representation of births and deaths of homology classes along time using bars. An example is shown in Table 1.

Finally, once persistence barcodes are obtained, persistent entropy can be computed.

Definition 5 (Persistent entropy [13]). *Given a filtered simplicial complex $\{K(t) : t \in F\}$, and the corresponding persistence barcode $B = \{a_i = [x_i, y_i) : i \in I\}$, the persistent entropy E of the filtered simplicial complex is calculated as follows:*

$$E = -\sum_{i \in I} p_i log(p_i)$$

where $p_i = \frac{l_i}{L}$, $l_i = y_i - x_i$, and $L = \sum_{i \in I} l_i$. In the case of an interval with no death time, $[x_i, \infty)$, the corresponding barcode $[x_i, m)$ will be considered, where $m = \max \{F\} + 1$.

The robustness of persistent homology to noise is guaranteed thanks to the following result, letting a stable comparison between signals.

Theorem 1 ([13]). *Given two functions, $f : V \rightarrow R$ and $g : V \rightarrow R$, defined on a set of vertices V of $\mathbb{R}^n$, then for every $\varepsilon > 0$, there exists $\delta > 0$ such that*

$$||f - g||_\infty \leq \delta \Rightarrow |E(f) - E(g)| \leq \varepsilon.$$

Machine Learning Background. Machine learning techniques are nowadays widely applied to solve classification problems.

A classification technique will use a 'training' dataset

$$D = \{ (\boldsymbol{v}_i, c_i) \mid \boldsymbol{v}_i \in \mathbb{R}^n, \ c_i \in \{0, \ldots, k\}, \ i \in \{1, \ldots, m\} \}$$

where $\{0, \ldots, k\}$ are the different possible classes. From this dataset, the classification algorithm will produce a classification model. This model can lately be applied to new inputs in order to predict the corresponding classes. There exist several classification techniques in machine learning. In our case, we focus our attention on support vector machine (see [1, 3, 4] and [6, Chap. 5]).

Table 1. Left: lower-star filtration. Right: the associated persistence barcode of the corresponding lower-star filtration pictured on the left.

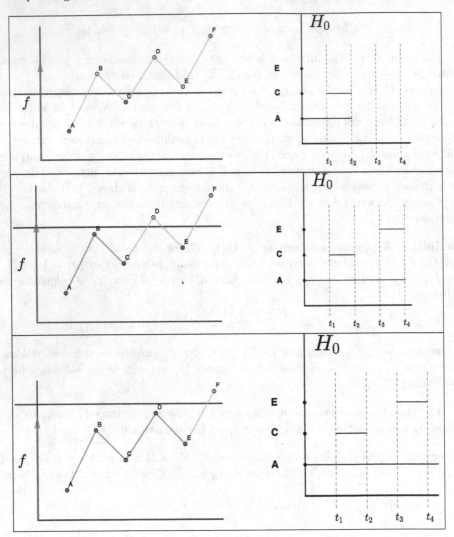

A support vector machine is a supervised learning technique that construct a hyperplane, driven by a linear function $b + \sum_{i=1}^{m} \alpha_i v_i^T v_i$, or a set of them that can be used to classify data. When this data is not linearly separable, a kernel trick is applied: the space is mapped to higher dimensions using a kernel function, $k(v, v') = \phi(v)^T \cdot \phi(v')$. Therefore, a support vector machine just creates hyperplanes that work as decision boundaries for classification after applying a deformation of the dataset in order to get a linearly separable representation. Then, formally, a support vector machine within a kernel makes predictions using the following function:

$$f(v) = b + \sum_{i=1}^{m} \alpha_i k(v, v_i)$$

where α is a vector of coefficients, k the kernel and b is a bias term. Finally, the coefficients are chosen as a result of an optimization problem of the separation margin between classes. Different kernel-based functions can be used, for example:

Kernels	
Linear	$k(v, v') = v^T \cdot v'$
Polynomial of degree d	$k(v, v') = (v^T \cdot v' + c)^d$
Gaussian	$k(u, v) = \mathcal{N}(u - v; 0, \sigma^2 I)$

where $\mathcal{N}(v; \mu, \Sigma)$ is the standard normal density.

Performance Metrics. Basically, we are dealing with a classification problem. Therefore, our main metric will be the **accuracy**, considered as the percentage of well classified data in a dataset:

$$\text{Accuracy} = \frac{m}{n}$$

where m is the number of well-classified data and n is the size of the full dataset used in the test.

Statistical Tool. The correlation coefficient of two random variables is a measure of their linear dependence. One correlation coefficient largely known and applied is the Pearson's correlation coefficient [11]:

$$\text{Pearson's correlation coefficient } \rho(A, B) = \frac{cov(A, B)}{\sigma_A \sigma_B}$$

where $cov(A, B)$ is the covariance and σ the standard deviation.

3   Methodology

As was previously anticipated, the shape of the wave of a speech signal can be meaningful to emotional speech recognition. Roughly speaking, we will compute persistent entropy to the lower-star filtration of the raw signal and then, we classify the signals by comparing these numbers using a support vector machine.

Let us now explain in details the methodology applied in this paper:

Step 1. Subsampling of the Signal. The size of each signal is reduced in order to face the complexity of the persistent homology algorithm. Besides, every signal of the dataset needs to be subsampled into the same size in order to fulfill the assumptions of Theorem 1. For example, we subsampled the signal pictured in Fig. 1 from 196997 points to 10000. The subsampling process was done uniformly on the signal, maintaining its shape and main distribution of the spikes. Furthermore, the experiments of Sect. 4 were also done using the dataset without subsampling reaching similar results. Then, we could assert that this type of subsampling does not loose relevant information for this approach.

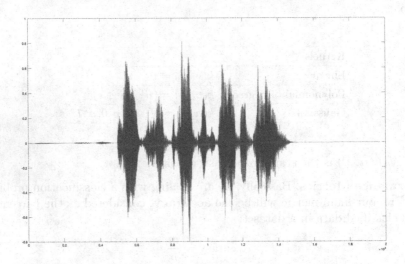

Fig. 1. Raw signal intensity graph of an angry emotion interpreted by the actor number 1 of the RAVDESS dataset.

Step 2. Introduction of Imperceptible Noise. Signals are slightly perturbed to fulfill the requirement of lower-start filtrations (see Definition 2): two points in the signal can not have the same height.

Step 3. Persistence Barcode Computation. The lower-star filtration technique is applied to the signals generated in Step 2, obtaining the associated persistence barcode. For example, the barcode associated to the signal of Fig. 1 can be seen in Fig. 2.

Step 4. Persistent Entropy Computation. Persistent entropy is computed applying the formula given in Definition 5 to the persistence barcodes obtained in Step 3.

Step 5. Support Vector Machine Classification. This step consists of the application of several support vector machines with different kernels in order to infer results and develop a classification predictor to emotions. The different possible kernels, previously introduced in the paper, are tested and the one with better accuracy is chosen.

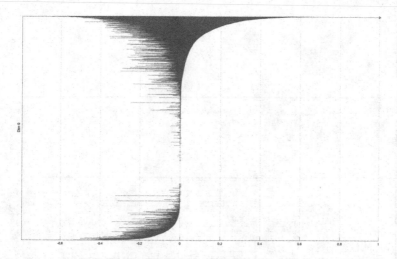

Fig. 2. Barcode of the signal shown in Fig. 1. The horizontal axis represents time. Every horizontal (blue) line represents the life of a 0-dimensional homology class. (Color figure online)

4 Experiments

The work-flow presented in the previous section was applied to the RAVDESS dataset [9]. This dataset is composed by 24 actors interpreting 60 audios each on different emotions and different intensity. Concretely, there are 4 audios for the neutral emotion and 8 audios for each of the seven remaining emotions. Consequently, there are 1440 different audios.

In Fig. 3, a box-plot of the persistent entropy of the 1440 audios grouped by the different emotions can be seen. We can infer that persistent entropy values vary depending on both the emotion and the person. It seems that there exists characteristic personal values and the range of every emotion can be really wide. For example, the persistent entropy values of the audio number 20 in Fig. 3, that is an example of happiness, varies from 5.1713 to 0.6923 depending on the person. Besides, the existing overlapping between the boxes tells us that emotions can not be distinguished from the rest by just the persistent entropy values of every script as a feature. This failure approximation is illustrated and explained in Experiment 1. However, some emotions can be differentiated by pairs even with this 'naive' approximation.

One thing that appealed our attention is the visual correlation that persistent entropy values tend to have per sexes as shown in Figs. 4 and 5. Even if the range is lower or higher depending on the person, in general, the peaks appear on the same places. To illustrate it, let us consider the correlation matrix between persistent homology values of the 60 audios grouped in the ones belonging to females and the ones belonging to males. We obtain that persistent entropy values are moderately correlated between same sex audios and badly correlated between different sexes (see Table 2). We think that it could be interesting the use of

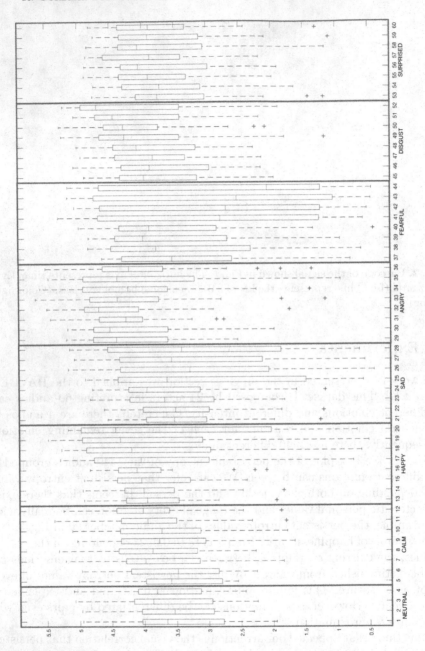

Fig. 3. (90° rotated figure) Horizontal axis represents the different 60 audios. Vertical axis represents persistent entropy. The big (red) rectangle clusters encloses persistent entropies of the audios per emotion (the respective emotion is indicated in the horizontal axis. The small (blue) rectangles are quartiles for the persistent entropy values. The vertical (blue) dashed lines mean the range of values of persistent entropy values. The (red) points are outliers. The horizontal (red) small lines are the mean persistent entropy value for the corresponding audio. (Color figure online)

more sophisticated measures of similarity apart from correlation. Furthermore, correlation results give us clues to the need of developing emotion classification within the dataset separated by sexes to reach better classification accuracy. Besides, we consider that persistent entropy values could even be a nice approach to people identification and not just to emotion recognition. However, this approach is far from the scope of this paper and its preliminary nature.

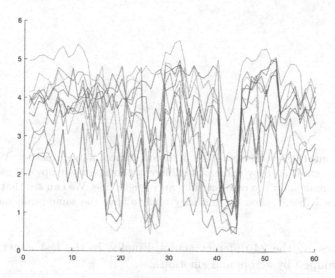

Fig. 4. Horizontal axis represents the different audios of actresses. Vertical axis represents persistent entropy value. The different persistent entropy values for the 60 audios of the same actress are connected by an straight line. We can see that shapes are correlated (see Table 2), showing that they tend to have the same peaks and downs.

Table 2. Mean values for the correlation coefficients of the entropy values grouped by sexes.

	Male actor	Female actor
Male actor	0.43	0.23
Female actor	0.23	0.49

In all the following experiments we use as the classification technique a support vector machine with fold cross validation and the kernel that provides the better accuracy from the ones explained previously. The training dataset will be the 1440 persistent entropy values grouped by different ways trying to get the features needed to reach our goal. In the first experiment we try the brute force approach using every script as a point of the training dataset. Then, in the second experiment, every point correspond to an emotion within its 24 persistent

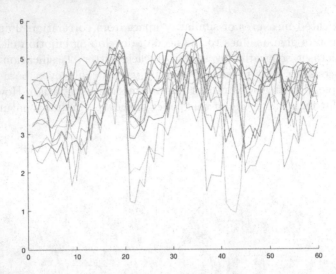

Fig. 5. Horizontal axis represents the different audios of male actors. Vertical axis represents persistent entropy value. The different persistent entropy values for the 60 audios of the same actor are connected by an straight line. We can see that shapes are correlated (see Table 2), showing that they tend to have the same peaks and downs.

entropy values by the 24 different actors. Finally, in the last experiment, the dataset is grouped by actors and emotions.

Experiment 1: Each persistent entropy value will be a point of the training dataset. In this case, 20.3% of accuracy is reached within a linear kernel. Some conclusions can be pointed out from this failed approach: The emotion recognition problem is a multidimensional one, in the sense that a 1-dimensional embedding is not enough to an acceptable classification result. Furthermore, this was anticipated by the overlapping of the different boxes at the box-plot of persistent entropy values showed in Fig. 3. Besides, the non correlation between persistent entropy values per sexes is a matter not taken into account in this experiment.

Experiment 2: Each point of the dataset is a vector of 24 features which correspond to the persistent entropy value of the same emotion interpreted by the 24 different actors. The dataset was separated in 40 points for training dataset and 20 points for test dataset and a gaussian kernel was used. Then, 92.5% of accuracy was reached on the training dataset and 90% on the test dataset. Furthermore, 96.66% accuracy was obtained on the full dataset. In our opinion, this experiment presents two main drawbacks. The first one is its difficult applicability as it needs 24 features of every emotion. However, withing long audio recordings, it could be cut into pieces and obtain enough features to classify. The other drawback is the small dataset we have for this experiment because of the way it has been grouped.

Experiment 3: In this experiment, each point of the dataset consists of a vector of 8 features, corresponding each feature to the persistent entropy value of the same emotion interpreted by the same actor. By this, the following accuracy Table 3 for classification by pair of emotions was obtained using a second degree polynomial kernel. Considering other results in the literature like [12] where 71% of accuracy was reached using Artificial Neural Networks, our results are really promising. However, we are still far from the 83% of accuracy reached in [19] using a multi-task hierarchical model. But we can say that, with just a first approximation, we could reach similar accuracy than those that already exist in the literature. Furthermore, as we are considering here just intensity and one type of filtration, only some features that characterize emotions are taken into account. Then, it gives us a nice starting point in order to improve the model by using different features of the signal and different filtrations.

Table 3. Prediction accuracy from pair of emotions using different support vector machine within different kernels.

Feelings	Calm	Happy	Sad	Angry	Fearful	Disgust	Surprised
Calm		77.1%	68.8%	81.2%	79.2%	72.9%	60.4%
Happy			62.5%	64.6%	60.4%	58.3%	64.6%
Sad				75%	62.5%	70.8%	60.4%
Angry					68.8%	77.1%	70.8%
Fearful						72.9%	72.9%
Disgust							75%
Surprised							

5   Conclusions and Future Work

A persistent entropy application has been developed in order to extract information from raw audio signals and solve a classification problem using support vector machine. Furthermore, a descriptive analysis of the computed persistent entropy values has been done, bringing up the characteristic values that exist by person and the existence of moderate correlation between persistent entropy values of emotions of people of the same sex. Additionally, we have provided insights showing that separating the dataset by sexes would get better accuracy for the classification task. Finally, three different experiments have been proposed: two of them can be considered successful. This makes evidence that topological data analysis tools are a nice approach to this task, being interesting the development of more sophisticated algorithms.

In this first approximation just β_0 has been used. However, there exists different processing techniques to signals that can obtain images from them and that would allow us to consider higher dimensional topology features that can be meaningful for the emotion recognition task. We could combine them to reach a better prediction skill.

Another interesting approach is training the machine learning classification tool with the audios interpreted by just one actor, obtaining a personal trained emotion predictor. However, RAVDESS dataset is not big enough to obtain interesting conclusions within this approach. Therefore, this would be a nice future work, in these days that it is quite easy to obtain lot of data from users.

Fig. 6. Landmarks points of one frame of a video of the RAVDESS dataset.

Furthermore, as the associated videos of the audios are available in the RAVDESS dataset, we would like to use the landmarks (see Fig. 6) as input to topological data analysis tools (like a Vietoris-Rips filtration) and combine this information within the one provided by the audios used in this paper. Similarly, one of the most relevant conclusions that KRISTINA project reached was that the combination of visual and audio features can develop better predictions than using them separately.

Acknowledgments. This research has been partially supported by MINECO, FEDER/UE under grant MTM2015-67072-P. We thank the anonymous reviewers for their valuable comments.

References

1. Boser, B.E., Guyon, I.M., Vapnik, V.N.: A training algorithm for optimal margin classifiers. In: COLT 1992, pp. 144–152. ACM, New York (1992)
2. Bredon, G.: Topology and Geometry. Springer, New York (1993). https://doi.org/10.1007/978-1-4757-6848-0
3. Cortes, C., Vapnik, V.: Support-vector networks. Mach. Learn. **20**(3), 273–297 (1995). https://doi.org/10.1023/A:1022627411411
4. Cristianini, N., Shawe-Taylor, J.: An Introduction to Support Vector Machines and Other Kernel-Based Learning Methods, 1st edn. Cambridge University Press, Cambridge (2000)
5. Edelsbrunner, H., Harer, J.L.: Computational Topology, an Introduction. American Mathematical Society, Providence (2010)

6. Geron, A.: Hands-on Machine Learning with Scikit-Learn and TensorFlow: Concepts, Tools, and Techniques to Build Intelligent Systems. O'Reilly Media, Sebastopol (2017)
7. Globerson, E., Amir, N., Golan, O., Kishon-Rabin, L., Lavidor, M.: Psychoacoustic abilities as predictors of vocal emotion recognition. Atten. Percept. Psychophys. **75**(8), 1799–1810 (2013)
8. Howard, D.M., Angus, J.: Acoustics and Psychoacoustics, 2nd edn. Butterworth-Heinemann, Newton (2000)
9. Livingstone, S.R., Russo, F.A.: The Ryerson audio-visual database of emotional speech and song (RAVDESS): a dynamic, multimodal set of facial and vocal expressions in North American English. PLOS ONE **13**(5), 1–35 (2018)
10. Ortony, A., Turner, T.J.: What's basic about basic emotions? Psychol. Rev. **97**(3), 315 (1990)
11. Pearson, K.: Note on regression and inheritance in the case of two parents. Proc. R. Soc. Lond. **58**, 240–242 (1895)
12. Popova, A.S., Rassadin, A.G., Ponomarenko, A.A.: Emotion recognition in sound. In: Kryzhanovsky, B., Dunin-Barkowski, W., Redko, V. (eds.) NEUROINFORMATICS 2017. SCI, vol. 736, pp. 117–124. Springer, Cham (2018). https://doi.org/10.1007/978-3-319-66604-4_18
13. Rucco, M., et al.: A new topological entropy-based approach for measuring similarities among piecewise linear functions. Signal Process. **134**, 130–138 (2017)
14. Russell, J.: A circumplex model of affect. J. Pers. Soc. Psychol. **39**(6), 1161–1178 (1980)
15. Schuller, B., Batliner, A.: Computational Paralinguistics: Emotion, Affect and Personality in Speech and Language Processing. Wiley, Hoboken (2013)
16. Ververidis, D., Kotropoulos, C.: Emotional speech recognition: resources, features, and methods. Speech Commun. **48**, 1162–1181 (2006)
17. Wasserman, L.: Topological data analysis. Ann. Rev. Stat. Appl. **5**(1), 501–532 (2018)
18. Yang, B., Lugger, M.: Emotion recognition from speech signals using new harmony features. Signal Process. **90**(5), 1415–1423 (2010). Special Section on Statistical Signal & Array Processing
19. Zhang, B., Essl, G., Mower Provost, E.: Recognizing emotion from singing and speaking using shared models, September 2015. https://doi.org/10.1109/ACII.2015.7344563
20. Zomorodian, A., Carlsson, G.: Computing persistent homology. Discret. Comput. Geom. **33**(2), 249–274 (2005)

Harmonic Holes as the Submodules of Brain Network and Network Dissimilarity

Hyekyoung Lee[1]([✉]), Moo K. Chung[4], Hongyoon Choi[1], Hyejin Kang[2],
Seunggyun Ha[1], Yu Kyeong Kim[3], and Dong Soo Lee[1,2]

[1] Seoul National University Hospital, Seoul, Republic of Korea
hklee.brain@gmail.com
[2] Seoul National University, Seoul, Republic of Korea
[3] SMG-SNU Boramae Medical Center, Seoul, Republic of Korea
[4] University of Wisconsin-Madison, Madison, WI, USA

Abstract. Persistent homology has been applied to brain network analysis for finding the shape of brain networks across multiple thresholds. In the persistent homology, the shape of networks is often quantified by the sequence of k-dimensional holes and Betti numbers. The Betti numbers are more widely used than holes themselves in topological brain network analysis. However, the holes show the local connectivity of networks, and they can be very informative features in analysis. In this study, we propose a new method of measuring network differences based on the dissimilarity measure of harmonic holes (HHs). The HHs, which represent the substructure of brain networks, are extracted by the Hodge Laplacian of brain networks. We also find the most contributed HHs to the network difference based on the HH dissimilarity. We applied our proposed method to clustering the networks of 4 groups, normal controls (NC), stable and progressive mild cognitive impairment (sMCI and pMCI), and Alzheimer's disease (AD). The results showed that the clustering performance of the proposed method was better than that of network distances based on only the global change of topology.

Keywords: Topological data analysis · Brain network
Alzheimer's disease · Harmonic holes · Hodge Laplacian

1 Introduction

Persistent homology has been widely applied to brain network analysis for finding the topology of networks at multiple scales [5,16,23,25]. Since a 'simplicial complex' is not a familiar term in brain network analysis, we refer to it as a 'network' that is generally used. It quantifies the shape of brain networks by using k-dimensional holes and their cardinality, the kth Betti number [2,10]. A persistence diagram (PD) summarizes the change of Betti numbers during the filtration of networks by recording when and how holes appear and disappear during the filtration. The persistent homology also provides distances for

© Springer Nature Switzerland AG 2019
R. Marfil et al. (Eds.): CTIC 2019, LNCS 11382, pp. 110–122, 2019.
https://doi.org/10.1007/978-3-030-10828-1_9

distinguishing networks such as the bottleneck distance and kernel-based distances [10,20]. Such distances mostly find network differences in their PDs. The Betti numbers and PDs are more often used than holes themselves in network applications.

Holes represent the submodule of brain networks. 0-dimensional holes, i.e., connected components, modules or clusters have been widely studied for finding functional or structural submodules in a brain [6,16,27]. On the other hand, 1-dimensional holes have been rarely used for brain network analysis [7,15,17, 19,24,26]. Most studies in brain network analysis do not use 2- and higher order simplexes in networks since networks. Therefore, all cycles in a network are considered as 1-dimensional holes. There are few network measures based on cycles in brain network analysis such as cycle probability and the change of the number of cycles during graph filtration [7,26]. These measures helped to compare the global property of networks but could not find the discriminative substructures of networks.

If higher order simplexes are introduced in a network, the number of 1-dimensional holes is significantly reduced due to the removal of filled-in triangles. The previous brain network studies that studied higher order simplexes mostly found holes based on Zomorodian and Carlsson's (ZC) algorithm [19,24,28]. The ZC algorithm is very fast in linear-time, however, it finds the sparse representation of a hole that identifies only one path around the hole and ignores the other paths. This introduces an ambiguity in hole identification in practice. A better approach would be to localize the holes by the eigen-decomposition of Hodge Laplacian of a network. Such holes are called as the harmonic holes (HHs). The HH shows all possible paths around the hole with their weights [12,13,18]. The HHs have been applied to brain network analysis for localizing persistent holes [15,17]. The 1-dimensional holes in a network with higher order simplexes have at least one indirect path between every two nodes. Thus, the holes are related to the abnormality or inefficiency of the network. The previous studies found the persistent holes with long duration in a network as abnormal holes, and localized them by harmonic holes. Therefore, the duration of holes was used instead of HHs in network discrimination.

In this paper, we propose a new measure for estimating network dissimilarity based on persistent HHs (HH dissimilarity). The proposed HH dissimilarity is motivated from the bottleneck distance. The bottleneck distance first estimates the correspondence between holes between networks that are represented by points in PDs, and then chooses the maximum among all the distances between the estimated pairs of holes [8]. The HH dissimilarity also estimates the correspondence between HHs of two different networks that are represented by real-valued eigenvectors, and takes the averaged dissimilarities of the estimated pairs of HHs. The advantage of HH dissimilarity is not only to measure the network differences but also to quantify a HH's contributions to the network differences. We will call the amount of HH's of contribution the *citation* of HH. This allows us to identify the discriminative subnetworks of networks.

The proposed method is applied to metabolic brain networks obtained from the FDG PET dataset in Alzheimer's disease neuroimaging initiative (ADNI). The dataset consists of 4 groups: normal controls (NC), stable and progressive mild cognitive impairment (sMCI and pMCI), and Alzheimer's disease (AD). We generated 2400 networks by bootstrap, and compared the clustering performance with the existing network distances such as L_2-norm (L2) of the difference between distance matrices, Gromov-Hausdorff (GH) distance, Kolmogorov-Smirnov (KS) distance of connected components and cycles (KS_0 and KS_1), and bottleneck distance of holes [3,6–8,16]. The results showed that the HH dissimilarity had the superior clustering performance than the other distance measures, and comparing local connectivities could be more helpful to discriminating the progression of Alzheimer's disease.

2   Materials and Methods

2.1   Data Sets, Preprocessing, and the Construction of Metabolic Connectivity

We used FDG PET images in ADNI data set (http://adni.loni.usc.edu). The ADNI FDG-PET dataset consists of 4 groups: 181 NC, 91 sMCI, 77 pMCI, and 135 AD (Age: 73.7 ± 5.9, range 56.1–90.1). FDG PET images were measured 30 to 60 min and they were averaged over all frames. The voxel size in the images were standardized in $1.5 \times 1.5 \times 1.5$ mm resolution. The images were spatially normalized to Montreal Neurological Institute (MNI) space using statistical parametric mapping (SPM8, www.fil.ion.ucl.ac.uk/spm). The details of data sets and preprocessing are given in [4]. The whole brain image was parcellated into 94 regions of interest (ROIs) based on automated anatomical labeling (AAL2) excluding cerebellum [21]. The 94 ROIs served as network nodes and their measurements were obtained by averaging FDG uptakes in the ROI. The averaged FDG uptake was globally normalized by the sum of 94 averaged FDG uptakes. The distance between 2 nodes was estimated by the diffusion distance on positive correlation between the measurements. The diffusion distance considers an average distance of all direct and indirect paths between 2 nodes via random walks [9]. The diffusion distance is known to be more robust to noise and outliers.

2.2   Harmonic Holes

Simplicial Complex. The algebraic topology extends the concept of a graph further to a simplicial complex. Suppose that a non-empty node set V is given. If the set of all subsets of V is denoted by 2^V, an abstract simplicial complex K is a subset of 2^V such that (1) $\emptyset \in K$, and (2) if $\sigma \in K$ and $\tau \in \sigma$, $\tau \in K$ [10,11]. Each $\sigma \in K$ is called a simplex. A i-dimensional simplex is an element with $i+1$ nodes, $v_1, ..., v_{i+1} \in V$, denoted by $\sigma_i = [v_1, ..., v_{i+1}]$. The dimension of K, denoted as $\dim K$, is the maximum dimension of a simplex $\sigma \in K$. The collection of σ_i's in K is denoted by K_i ($-1 \leq i \leq \dim K$). The number of simplices in

K_i is denoted as $|K_i|$. The i-skeleton of K is defined as $K^{(i)} = K_0 \cup \cdots \cup K_i$ $(0 \le i \le \dim K)$. Thus, a graph with nodes and edges is 1-skeleton $K^{(1)}$. In this paper, we will only consider 2-skeleton $K^{(2)}$ of a simplicial complex that includes nodes, edges, and triangles. For convenience, we call it *a (simplicial) network* [14].

Incidence Matrix. We denote a $|K_i|$-dimensional integer space as $\mathbb{Z}^{|K_i|}$. Given a finite simplicial complex K, a chain complex C_i is defined in $\mathbb{Z}^{|K_i|}$ [10,28]. The boundary operator ∂_i and coboundary operator $\partial_i^\top$ for $i = 1, \ldots, N$ $(N > 0)$ are functions such that $\partial_i : C_i \to C_{i-1}$ and $\partial_i^\top : C_{i-1} \to C_i$, respectively. We define $\partial_i = 0$ for $i < 1$ or $i > N$.

Given $\sigma_i = [v_1, ..., v_{i+1}] \in C_i$, the boundary of σ_i is algebraically defined as

$$\partial_i \sigma_i = \sum_{j=1}^{i+1} (-1)^{j-1} [v_1, \ldots, v_{j-1}, v_{j+1}, \ldots, v_{i+1}].$$

If the sign of σ_{i-1} in $\partial_i \sigma_i$ is positive/negative, it is called positively/negatively oriented with respect to σ_i. We denote the positive/negative orientation by $\sigma_{i-1} \in_{+/-} \sigma_i$. The boundary of the boundary is always zero, i.e., $\partial_{i-1}\partial_i = 0$.

If the simplicial complex K has

$$K_i = \left\{ \sigma_i^1, \cdots, \sigma_i^{|K_i|} \right\}, \quad K_{i-1} = \left\{ \sigma_{i-1}^1, \cdots, \sigma_{i-1}^{|K_{i-1}|} \right\},$$

the boundary operator ∂_i is represented by the ith incidence matrix $\boldsymbol{M}_i \in \mathbb{Z}^{|K_{i-1}| \times |K_i|}$ such that [12,13,18]

$$[\boldsymbol{M}_i]_{mn} = \begin{cases} 1 & \text{if } \sigma_{i-1}^m \in_+ \sigma_j^n, \\ -1 & \text{if } \sigma_{i-1}^m \in_- \sigma_j^n, \\ 0 & \text{otherwise.} \end{cases} \tag{1}$$

The coboundary operator $\partial_i^\top$ is represented by $\boldsymbol{M}_i^\top$. σ_i^n in K_i is represented by a vector in $\mathbb{Z}^{|K_i|}$ in which the nth entry is 1 and the rest is 0. The linear combination of σ_i's can be represented by the linear combination of $|K_i|$-dimensional vectors.

Combinatorial Hodge Laplacian. A combinatorial Hodge Laplacian $\boldsymbol{L}_i : C_i \to C_i$ is defined by

$$\boldsymbol{L}_i = \boldsymbol{L}_i^{up} + \boldsymbol{L}_i^{down} = \boldsymbol{M}_{i+1}\boldsymbol{M}_{i+1}^\top + \boldsymbol{M}_i^\top \boldsymbol{M}_i, \tag{2}$$

where $\boldsymbol{L}_i^{up} \in \mathbb{Z}^{|K_i| \times |K_i|}$ and $\boldsymbol{L}_i^{down} \in \mathbb{Z}^{|K_i| \times |K_i|}$ are composite functions $\partial_{i+1}\partial_{i+1}^\top : C_i \to C_{i+1} \to C_i$ and $\partial_i^\top \partial_i : C_i \to C_{i-1} \to C_i$, respectively [12–14,18] The kernel and image of $\boldsymbol{L}_i$ are denoted by $\ker \boldsymbol{L}_i$ and $\text{img} \boldsymbol{L}_i$, respectively. The $\ker \boldsymbol{L}_i$ is called harmonic classes H_i [14].

The ith homology and cohomology groups of $C = \{C_i, \partial_i\}$ are defined respectively by

$$\tilde{H}_i(C) = \ker \partial_i / \text{img} \partial_{i+1} \quad \text{and} \quad \tilde{H}^i(C) = \ker \partial_{i+1}^\top / \text{img} \partial_i^\top.$$

Theorem 1 (Combinatorial Hodge Theory [12,14,18]). *Suppose that a chain complex* $\{C_i(X;\mathbb{R}), \partial_i\}$ *is given for* $i = 0, \ldots, N$, *and* $C_i(X;\mathbb{R})$ *is considered as an* $\mathbb{R}$-*vector space. Harmonic classes* H_i *obtained by the combinatorial Laplacian* L_i *are congruent to the* ith *homology and cohomology groups,* $\tilde{H}_i$ *and* $\tilde{H}^i$ *of* C, *i.e.,*

$$H_i \cong \tilde{H}_i(C;\mathbb{R}) \cong \tilde{H}^i(C;\mathbb{R}).$$

Proof. $\mathrm{rank} H_i = \mathrm{rank} C_i - \mathrm{rank} L_i = \mathrm{rank} C_i - (\mathrm{rank} \partial_i + \mathrm{rank} \partial_{i+1}) = \mathrm{rank} \tilde{H}_i(C;\mathbb{R})$.

The harmonic classes $H_i = \ker L_k$ is also called a harmonic space [14]. The homology group $\tilde{H}_i$ in persistent homology can be replaced with a harmonic space H_i, and the rank of H_i is the same as the ith Betti number. We call a hole in H_i a harmonic hole (HH), and a hole in $\tilde{H}_i$ estimated by Smith normal form a binary hole [28].

Given a simplicial network with p nodes, q edges, and r filled-in triangles, we estimate $L_1 \in \mathbb{Z}^{q \times q}$ in (2), and $H_i = \{x \in \mathbb{R}^{q \times 1} | L_1 x = 0\}$. The eigenvector of L_1 with zero eigenvalue, $x \in \mathbb{R}^{q \times 1}$ represents a HH. The entry of x can be positive or negative depending on the orientation of edges in the hole. The absolute value of the entry of x represents the weight of the corresponding edge in the hole. Since x and $-x$ have zero eigenvalue, they represent the same hole, and $\| x \| = 1$.

Computing Persistent HHs. In this study, we have the distances between pairs of nodes in a brain network. Given a set of nodes and their distances, Rips complex with threshold ϵ is the clique complex induced by a set of edges with their distances less than ϵ. Rips filtration is the nested sequence of Rips complexes obtained by increasing threshold ϵ. To compute persistent holes over threshold, we perform Rips filtration on brain network nodes [2,10].

Zomorodian and Carlsson developed an efficient algorithm for computing persistent holes based on the Smith normal form [28]. It is an incremental algorithm that updates the range and null spaces of incidence matrices during Rips filtration. The representation of a persistent binary hole is changed by adding simplexes during Rips filtration. The ZC algorithm chose the youngest binary hole at the birth of the persistent hole. The ZC algorithm is fast in practically linear-time, however, the obtained binary hole shows only one path around the hole and the other paths are ignored. On the other hand, a HH shows all possible paths around the persistent hole, and represents the contribution of a path to the generation of the hole by edge weights in the path. Thus, the HH is better in localizing a persistent hole than a binary hole when we want to extract local connectivity in a brain network. However, there is no algorithm for estimating persistent HHs during the filtration in literature.

In this study, we will estimate the youngest persistent HHs just like the ZC algorithm. First, we sort edges $e_1, \ldots, e_q$ in the ascending order of an edge distance, and perform the Rips filtration by the fast ZC algorithm. To avoid having the same edge distance, we select the ordered index $1, \ldots, q$ as the filtration value, instead of the edge distance. The reason for performing the ZC algorithm first is that the computation of eigen-decomposition at every filtration value is too expensive. Then, we obtain a PD which is the set of the birth and death thresholds of persistent holes. If a persistent hole appears at i_X and disappears at i_Z, we perform the eigen-decomposition of Hodge Laplacian at i_X, i_Z, and $i_Y = i_Z - 1$ to estimate the corresponding HH. The i_Y is the threshold just before the death of the persistent hole.

The harmonic spaces at i_X, i_Y, and i_Z are written by matrices

$$H_X = [\boldsymbol{x}_1, \cdots, \boldsymbol{x}_l] \in \mathbb{R}^{q \times l}, H_Y = [\boldsymbol{y}_1, \cdots, \boldsymbol{y}_m] \in \mathbb{R}^{q \times m}, H_Z = [\boldsymbol{z}_1, \cdots, \boldsymbol{z}_n] \in \mathbb{R}^{q \times n},$$

respectively. The HH appearing at i_X and disappearing at i_Z will be in H_X and H_Y, but not in H_Z. We find which $\boldsymbol{y} \in H_Y$ does not depend on $\boldsymbol{z}_i$'s in H_Z. If $\boldsymbol{y} \in H_Y$ depends on H_Z, the smallest singular value of the matrix $[H_Z, \boldsymbol{y}]$ is close to 0. It implies that $\boldsymbol{y}$ still exists in H_Z. Therefore, we choose $\boldsymbol{y} \in H_Y$ such that

$$\boldsymbol{y} = \arg \max_{\boldsymbol{y} \in H_Y} \{\text{the smallest singular value of } [H_Z, \boldsymbol{y}]\}. \tag{3}$$

The chosen $\boldsymbol{y}$ by (3) is the oldest persistent HH. Next, we choose the youngest persistent HH $\boldsymbol{x} \in H_X$ such that

$$\boldsymbol{x} = \arg \min_{\boldsymbol{x} \in H_X} \{\text{the smallest singular value of } [\boldsymbol{x}, \boldsymbol{y}]\} = \arg \min_{\boldsymbol{x} \in H_X} \{1 - |\boldsymbol{x}^\top \boldsymbol{y}|\}. \tag{4}$$

This procedure is repeated for all persistent holes. The incidence matrices are already estimated during the ZC algorithm. Since the incidence matrices and their combinatorial Hodge Laplacian are very sparse, the computation of persistent HHs is not so hard in our experiments. In our experiments, the total number of persistent holes during the filtration is not more than 50, and the number of persistent holes at each filtration value is not more than 20.

2.3   HH Dissimilarity

Bottleneck Distance. If K_a and K_b have m and n persistent holes. The PDs of K_a and K_b are denoted respectively by $PD_a = \{t_1^a, \cdots, t_m^a\}$ and $PD_b = \{t_1^b, \cdots, t_n^b\}$, where t_i is a point with the birth and death thresholds of the corresponding hole. Bottleneck distance between two simplicial complexes, K_a and K_b is defined by [8]

$$D_B(K_a, K_b) = d(PD_a, PD_b) = \inf_{\eta: PD_a \to PD_b} \sup_{t \in PD_a} \| \boldsymbol{t} - \eta(t) \|_\infty,$$

where η is a bijection from PD_a to PD_b and $\| (x, y) \|_\infty = \max \{|x|, |y|\}$ is the L_∞-norm. If there is no corresponding hole in the other PD because of $m \neq n$, the points on the diagonal line $x = y$ that have the shortest distance from the point $\boldsymbol{t}$ are included. In this way, the bottleneck distance measures network distance by the difference of the birth and death thresholds of holes, not by the difference between holes themselves.

Dissimilarity Between HHs. If the eigenvectors with zero eigenvalues of two different combinatorial Laplacians are denoted by x and y, their dissimilarity is defined by one minus the absolute value of their inner product, i.e.,

$$d_h(x, y) = 1 - |x^\top y|. \tag{5}$$

This is the smallest singular value of the matrix $[x, y]$ in (4) that shows the dependency between x and y. If x and y are similar, their dissimilarity is close to 0; otherwise, it is close to 1.

HH Dissimilarity. Suppose that two networks K_a and K_b have m and n persistent HHs, denoted by $H_a = [x_1^a, \cdots, x_m^a]$ and $H_b = [x_1^b, \cdots, x_n^b]$, respectively. The dissimilarity based on persistent HHs (HH dissimilarity) is defined by

$$D_H(K_a, K_b) = d(H_a, H_b) = \inf_{\zeta : H_a \to H_b} \frac{1}{\min(m, n)} \sum_{x \in H_a} d_h(x, \zeta(x)), \tag{6}$$

where ζ is a bijection from H_a to H_b.

The correspondence ζ between persistent HHs in two different networks is determined by minimizing the total distances between the pairs of HHs based on Munkres assignment algorithm, also known as Hungarian algorithm. Some of persistent HHs can not find their corresponding HHs in the other network because of $m \neq n$. In this study, we ignore them and average the dissimilarities of the obtained pairs of persistent HHs.

Citation of HH. The advantage of using HH dissimilarity is the ability to quantify how much a persistent HH contributes in differentiating networks. The degree of the contribution of HH is called the citation of HH. If a persistent HH x in H_a corresponds to a persistent HH $y = \zeta(x)$ in H_b in (6), their dissimilarity is $d_h(x, y) = 1 - |x^\top y|$, and their similarity is defined by $|x^\top y|$. If the persistent HHs of l networks are denoted by $\mathcal{H} = \{H_1, \cdots, H_l\}$ and they are compared with H_a, the citation of x is defined by

$$\sum_{\zeta(x) \in H, \forall H \in \mathcal{H}} |x^\top \zeta(x)|.$$

If we find the most cited HHs by comparing networks within a group, we can determine which submodule makes two networks in a group close to each other. Furthermore, if we find the most cited HHs by comparing network between groups, we can determine which submodule makes differences.

3   Results

3.1   Brain Network Construction

We had 4 groups, NC, sMCI, pMCI, and AD which had 181, 91, 77 and 135 subjects, respectively. The subjects in a group could be heterogeneous. Thus, we obtained 600 bootstrap samples from each group by randomly selecting the sub-set of the number of subjects in each group with replacement [22]. The number of bootstrap samples was heuristically determined in comparison with previous study [22]. We constructed 600 bootstrapped networks from bootstrap samples in each group by diffusion distance in Sect. 2.1. The total number of generated brain networks was 2400.

3.2 Network Clustering

We clustered 2400 bootstrapped brain networks into 4 groups by Ward's hierar-chical clustering method. The Ward's hierarchical clustering method found the group labels based on the distance between data points, which is a network in our application. The network distance was estimated by (a) L2, (b) GH distance, (c) KS_0, (d) KS_1, (e) bottleneck distance of holes, and (f) HH dissimilarity [3, 6–8, 16]. The obtained distance matrices of 2400 networks were shown in Fig. 1. After clustering networks, we matched the estimated group label with the true group label of networks and calculated the clustering accuracy of 8 distance matrices. The clustering accuracy of 8 distance matrices was shown in Table 1. We also clustered 1200 bootstrapped networks in sMCI and pMCI into 2 groups by the same way. The clustering accuracy was shown in Table 1.

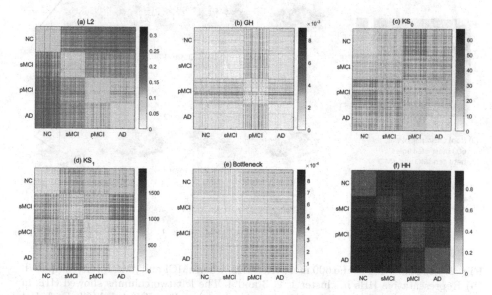

Fig. 1. Distance of 2400 networks. (a) L2, (b) GH, (c) KS_0, (d) KS_1, (e) Bottleneck, and (f) HH. The 2400 networks were sorted in the order of NC, sMCI, pMCI, and AD. Each group had 600 networks. The clustering accuracy is shown in Table 1.

Table 1. Clustering accuracy

Distance	4 groups (NC, sMCI, pMCI, and AD)	2 groups (sMCI and pMCI)
(a) L2	66.09%	98.50%
(b) GH	45.96%	87.58%
(c) KS_0	52.54%	74.00%
(d) KS_1	77.38%	79.83%
(e) Bottleneck	45.71%	76.58%
(f) HH	100%	100%

3.3 The Most Cited HHs

We selected the 600 most cited HHs within NC, sMCI, pMCI, and AD, and divided them into 5 clusters based on the dissimilarity between HHs in (5). In Fig. 3(a–d), because the dissimilarity of HHs in the cluster 5 was large, we considered HHs in the cluster 5 as outliers. We calculated the center of HHs in clusters 1, 2, 3, and 4, by selecting the HH with the minimum sum of dissimilar-

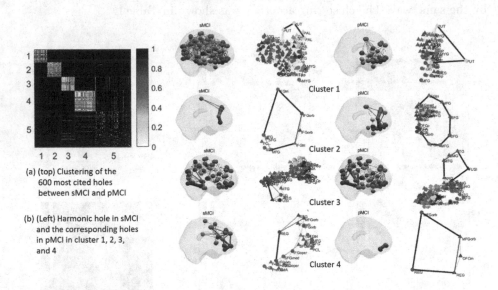

(a) (top) Clustering of the 600 most cited holes between sMCI and pMCI

(b) (Left) Harmonic hole in sMCI and the corresponding holes in pMCI in cluster 1, 2, 3, and 4

Fig. 2. (a) Clustering of the 600 most cited HHs when sMCI and pMCI were compared. (b) Representative HHs in cluster 1, 2, 3 and 4. The left two columns showed HHs in sMCI and the right two columns showed the corresponding HHs in pMCI. Each HH was visualized in a brain and in a 2-dimensional plane. The shape of the HH was more clearly shown in the plane, and the location of the HH could be checked in the brain. The color of nodes was determined by the location of nodes in a brain: frontal (red), parietal (blue), temporal (green), occipital (purple), subcortical (yellow), and limbic (orange) regions. If the edge weight was larger in a HH, the color of edge was darker and the width of edge was larger. (Color figure online)

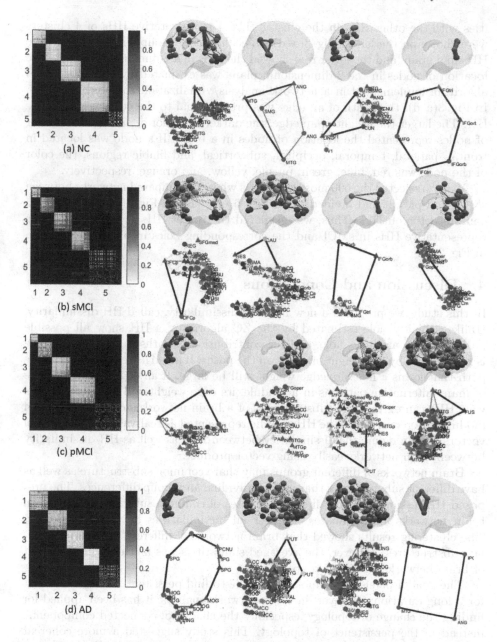

Fig. 3. Distance matrix of the 600 most cited HHs within (a) NC, (b) sMCI, (c) pMCI, and (d) AD. The most cited holes were clustered into 5 groups. The last cluster 5 had outliers with large dissimilarities between HHs. The representative HHs of the first 4 clusters were plotted on the right. The upper row showed the HHs in a brain and the lower row showed the HHs in a 2-dimensional plane. (Color figure online)

ities with the other HHs in the cluster. The 4 representative HHs of 4 clusters were shown on the left of Fig. 3(a–d). In each panel, the upper row showed the HHs in a brain, and the lower row showed the HHs in a 2-dimensional plane. The location of nodes in the 2-dimensional plane was estimated by Kamada-Kawai algorithm implemented in a network analysis/visualization toolbox, Pajek [1]. In Fig. 3(a–d), the width of an edge was proportional to the edge weight in the HH. The larger the weight of an edge, the darker the color of an edge. The color of nodes represented the location of nodes in a brain. If a node was located in frontal, parietal, temporal, occipital, subcortical, and limbic regions, the color of the node was red, blue, green, purple, yellow, and orange, respectively.

We also selected the 600 most cited HHs when we compared networks between sMCI and pMCI, and divided them into 5 clusters. In Fig. 2(a), the cluster 5 contained the outliers. Thus, we estimated the center HHs in cluster 1–4. The representative HHs in sMCI and the corresponding holes in pMCI were shown in Fig. 2(b).

4   Discussion and Conclusions

In this study, we proposed a new network dissimilarity, called HH dissimilarity. Unlike a binary hole estimated by the ZC algorithm, a HH show all possible paths of edges around a hole, and the contribution of paths to forming the hole is represented by the weight of edges on the paths. If an edge belongs to a unique path that forms a hole, its edge weight will be large. If an edge belongs to one of many alternative paths as in a module, its edge weight will be small. In this way, HHs can extract the substructures of a brain network including holes and modules. Moreover, since the HHs can be represented as real-valued orthonormal vectors we can define the dissimilarity between HHs as well as HH dissimilarity between brain networks easily using vector product.

Brain networks of different groups may share common substructure as well as have different substructures that make individual and group differences. The proposed HH dissimilarity first finds candidates of common substructures between brain networks and estimates the over all dissimilarities between candidates. The clustering results showed that brain networks of different groups had similar substructures, however, the averaged similarities was much larger than that of brain networks within a group.

The goal of persistent homology may be to find persistent features that last for a long duration. However, in brain network analysis, it has been applied for finding the change of topology, especially the change of connected components, instead of the persistence of topology. This study suggested a more coherent framework to observe, capture, and quantify the change of holes in brain networks. Depending on imaging modality and study populations, brain networks may have different characteristics of shapes. Therefore, it is necessary to apply proper network measures to brain networks depending on modality and population. The results showed that when the Alzheimer's disease progresses, the hole structure was changed in metabolic brain networks, and HHs and HH dissimilarity could predict the disease progression.

Acknowledgements. Data used in preparation of this article were obtained from the Alzheimer's Disease Neuroimaging Initiative (ADNI) database (adni.loni.usc.edu). As such, the investigators within the ADNI contributed to the design and implementation of ADNI and/or provided data but did not participate in analysis or writing of this report. A complete listing of ADNI investigators can be found at http://adni.loni.usc.edu. This work is supported by Basic Science Research Program through the National Research Foundation (NRF) (No. 2013R1A1A2064593 and No. 2016R1D1A1B03935463), NRF Grant funded by MSIP of Korea (No. 2015M3C7A1028926 and No. 2017M3C7A1048079), NRF grant funded by the Korean Government (No. 2016R1D1A1A02937497, No. 2017R1A5A1015626, and No. 2011-0030815), and NIH grant EB022856.

References

1. Batagelj, V., Mrvar, A.: Pajek - analysis and visualization of large networks. In: Jünger, M., Mutzel, P. (eds.) Graph Drawing Software. Mathematics and Visualization, pp. 77–103. Springer, Heidelberg (2003). https://doi.org/10.1007/978-3-642-18638-7_4

2. Carlsson, G., Collins, A., Guibas, L.J.: Persistence barcodes for shapes. Int. J. Shape Model. **11**, 149–187 (2005)

3. Carlsson, G., Ishkhanov, T., de Silva, V., Zomorodian, A.: On the local behavior of spaces of natural images. Int. J. Comput. Vis. **76**(1), 1–12 (2008)

4. Choi, H., Jin, K.H.: Predicting cognitive decline with deep learning of brain metabolism and amyloid imaging. Behav. Brain Res. **344**, 103–109 (2018). https://doi.org/10.1016/j.bbr.2018.02.017. https://www.sciencedirect.com/science/article/pii/S0166432818301013

5. Chung, M.K., Bubenik, P., Kim, P.T.: Persistence diagrams of cortical surface data. In: Prince, J.L., Pham, D.L., Myers, K.J. (eds.) IPMI 2009. LNCS, vol. 5636, pp. 386–397. Springer, Heidelberg (2009). https://doi.org/10.1007/978-3-642-02498-6_32

6. Chung, M.K., Villalta-Gil, V., Lee, H., Rathouz, P.J., Lahey, B.B., Zald, D.H.: Exact topological inference for paired brain networks via persistent homology. In: Niethammer, M., et al. (eds.) IPMI 2017. LNCS, vol. 10265, pp. 299–310. Springer, Cham (2017). https://doi.org/10.1007/978-3-319-59050-9_24

7. Chung, M.K., et al.: Topological brain network distances. arXiv:1809.03878 [stat.AP] (2018). https://arxiv.org/abs/1809.03878

8. Cohen-Steiner, D., Edelsbrunner, H., Harer, J.: Stability of persistence diagrams. Discrete Comput. Geom. **37**, 103–120 (2007)

9. Coifman, R.R., Lafon, S., Lee, A.B., Maggioni, M., Warner, F., Zucker, S.: Geometric diffusions as a tool for harmonic analysis and structure definition of data: diffusion maps. In: Proceedings of the National Academy of Sciences, pp. 7426–7431 (2005)

10. Edelsbrunner, H., Harer, J.: Persistent homology - a survey. Contemp. Math. **453**, 257–282 (2008)

11. Edelsbrunner, H., Harer, J.: Computational Topology: An Introduction. American Mathematical Society Press, New York (2009)

12. Friedman, J.: Computing Betti numbers via combinatorial Laplacians. In: Proceedings of the 28th Annual ACM Symposium on the Theory of Computing, pp. 386–391 (1996)

13. Horak, D., Jost, J.: Spectra of combinatorial Laplace operators on simplicial complexes. Adv. Math. **244**, 303–336 (2013)
14. Kim, Y.-J., Kook, W.: Harmonic cycles for graphs. Linear Multilinear Algebra, 1–11 (2018). https://doi.org/10.1080/03081087.2018.1440519
15. Lee, H., Chung, M.K., Kang, H., Choi, H., Kim, Y.K., Lee, D.S.: Abnormal hole detection in brain connectivity by kernel density of persistence diagram and Hodge Laplacian. In: 2018 IEEE 15th International Symposium on Biomedical Imaging (ISBI 2018), pp. 20–23, April 2018. https://doi.org/10.1109/ISBI.2018.8363514
16. Lee, H., Chung, M.K., Kang, H., Kim, B.N., Lee, D.S.: Persistent brain network homology from the perspective of dendrogram. IEEE Trans. Med. Imaging **31**, 2267–2277 (2012)
17. Lee, H., Chung, M.K., Kang, H., Lee, D.S.: Hole detection in metabolic connectivity of alzheimer's disease using k–laplacian. In: Golland, P., Hata, N., Barillot, C., Hornegger, J., Howe, R. (eds.) MICCAI 2014, LNCS, vol. 8675, pp. 297–304. Springer, Cham (2014). https://doi.org/10.1007/978-3-319-10443-0_38
18. Lim, L.H.: Hodge Laplacians on graphs. Geometry and topology in statistical inference. In: Proceedings of Symposia in Applied Mathematics, vol. 73 (2015)
19. Petri, G., et al.: Homological scaffolds of brain functional networks. J. Roy. Soc. Interface **11**(101), 20140873 (2014). https://doi.org/10.1098/rsif.2014.0873
20. Reininghaus, J., Huber, S., Bauer, U., Kwitt, R.: A stable multi-scale kernel for topological machine learning. In: The IEEE Conference on Computer Vision and Pattern Recognition (CVPR), pp. 4741–4748, June 2015
21. Rolls, E.T., Joliot, M., Tzourio-Mazoyer, N.: Implementation of a new parcellation of the orbitofrontal cortex in the automated anatomical labeling atlas. Neuroimage **122**, 1–5 (2015)
22. Sanabria-Diaz, G., Martìnez-Montes, E., Melie-Garcia, L., Alzheimer's Disease Neuroimaging Initiative: Glucose metabolism during resting state reveals abnormal brain networks organization in the Alzheimer's disease and mild cognitive impairment. PLOS ONE **8**(7), 1–25 (2013). https://doi.org/10.1371/journal.pone.0068860
23. Singh, G., Memoli, F., Ishkhanov, T., Sapiro, G., Carlsson, G., Ringach, D.L.: Topological analysis of population activity in visual cortex. J. Vis. **8**, 1–18 (2008)
24. Sizemore, A., Giusti, C., Kahn, A., Vettel, J., Betzel, R., Bassett, D.: Cliques and cavities in the human connectome. J. Comput. Neurosci. **44**, 115–145 (2018)
25. Solo, V., et al.: Connectivity in fMRI: blind spots and breakthroughs. IEEE Trans. Med. Imaging **37**(7), 1537–1550 (2018). https://doi.org/10.1109/TMI.2018.2831261
26. Sporns, O., Tononi, G., Edelman, G.: Theoretical neuroanatomy: relating anatomical and functional connectivity in graphs and cortical connection matrices. Cereb. Cortex **10**(2), 127–141 (2000). https://doi.org/10.1093/cercor/10.2.127
27. Sporns, O., Betzel, R.F.: Modular brain networks. Ann. Rev. Psychol. **67**, 19.1–19.28 (2016)
28. Zomorodian, A., Carlsson, G.: Computing persistent homology. Discrete Comput. Geom. **33**, 249–274 (2005)

Persistent 1-Cycles: Definition, Computation, and Its Application

Tamal K. Dey, Tao Hou$^{(\boxtimes)}$, and Sayan Mandal

Department of Computer Science and Engineering, The Ohio State University,
Columbus, OH, USA
{dey.8,hou.332,mandal.25}@osu.edu

Abstract. Persistence diagrams, which summarize the birth and death of homological features extracted from data, are employed as stable signatures for applications in image analysis and other areas. Besides simply considering the multiset of intervals included in a persistence diagram, some applications need to find representative cycles for the intervals. In this paper, we address the problem of computing these representative cycles, termed as *persistent 1-cycles*. The definition of persistent cycles is based on the interval module decomposition of persistence modules, which reveals the structure of persistent homology. After showing that the computation of the optimal persistent 1-cycles is NP-hard, we propose an alternative set of *meaningful* persistent 1-cycles that can be computed with an efficient polynomial time algorithm. We also inspect the stability issues of the optimal persistent 1-cycles and the persistent 1-cycles computed by our algorithm with the observation that the perturbations of both cannot be properly bounded. We design a software which applies our algorithm to various datasets. Experiments on 3D point clouds, mineral structures, and images show the effectiveness of our algorithm in practice.

Keywords: Persistent homology · Persistent cycle · Minimal cycle · NP-hardness · Image segmentation

1 Introduction

Persistent homology [18] is an important invention leading to Topological Data Analysis, where the associated persistence diagrams serve as stable signatures for various datasets [10] including the ones in image analysis [6,15]. Persistent homology has its theoretical foundations rooted in quiver theory [11], in which case any persistence module indexed by a finite subcategory of $\mathbb{R}$ can be decomposed into a direct sum of interval modules and the set of intervals of the interval modules, which constitute the persistence diagram, is unique for a persistence module [7].

Supported by NSF grants CCF-1740761 and CCF-1839252.

© Springer Nature Switzerland AG 2019
R. Marfil et al. (Eds.): CTIC 2019, LNCS 11382, pp. 123–136, 2019.
https://doi.org/10.1007/978-3-030-10828-1_10

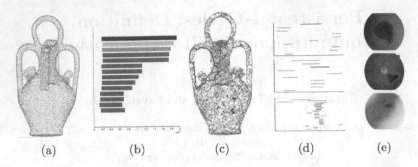

(a)   (b)   (c)   (d)   (e)

Fig. 1. (a) Point cloud of Botijo model. (b,c) Barcode and persistent 1-cycles for Botijo, where the 3 longest bars (dark blue, light blue, and green) have their corresponding persistent 1-cycles drawn with the same colors. (d,e) Barcode and persistent 1-cycles for the retinal image, with each green cycle corresponding to a red bar. (Color figure online)

Besides simply incorporating the persistence diagrams, some applications bring about the need of finding representative cycles for persistent homology [19,26]. The computation of representative cycles for homology groups with $\mathbb{Z}_2$ coefficients has been extensively studied over the decades. While a polynomial time algorithm computing an optimal basis for first homology group H_1 [14] has been proposed, finding an optimal basis for dimension greater than one and localizing a homology class of any dimension are proved NP-hard [9]. There are a few works addressing the problem of finding representatives for persistent homology, some of which compute an optimal cycle at the birth index of an interval but do not consider what actually die at the death index [19,20]. Obayashi [23] formalizes the computation of optimal representatives for a finite interval as an integer programming problem. He advocates solving it with linear programs though the correctness is not necessarily guaranteed. Wu et al. [26] proposed an algorithm for computing an optimal representative for a finite interval with a worst-case complexity exponential to the cardinality of the persistence diagram.

In this paper, we study the problem of computing representative cycles for persistent first homology group (H_1-persistent homology) with $\mathbb{Z}_2$ coefficients. We term theses cycles as *persistent 1-cycles* and show that the computation of the optimal cycles is NP-hard. Then, we propose an alternative set of *meaningful* persistent 1-cycles with an efficient polynomial time algorithm. Specifically, as interval module decomposition reveals the structure of persistence modules, we define persistent cycles which fit into this structure directly. Although similar definitions for finite intervals have already been proposed [23,26], to our knowledge, explicit explanation of how the representative cycles are related to persistent homology has not been addressed. Furthermore, we inspect the stability of the minimal persistent 1-cycles and persistent 1-cycles computed by our algorithm. The perturbations of both classes of cycles turn out to be unstable. So, in this regard, our polynomial time algorithm is not any worse than an optimal cycle generating algorithm though is much more efficient in terms of the time complexity.

We use a software based on our algorithm to generate tight persistent 1-cycles on 3D point clouds and 2D images as shown in Fig. 1. We experiment with various datasets commonly used in geometric modeling, computer vision and material science, details of which are given in Sect. 6. The *software, named* PersLoop, *along with an introductory video* and other supplementary materials are available at the project website http://web.cse.ohio-state.edu/~dey.8/PersLoop.

2 Background

In this paper, we adopt the categorical definition of persistence module [4]. A category C consists of objects and morphisms from an object to another object. A functor $F : C \to B$ from C to another category B is a mapping such that any object c of C is mapped to an object $F(c)$ of B and any morphism $f : c \to c'$ of C is mapped to a morphism $F[f] : F(c) \to F(c')$ of B. We recommend [1] for the exact definitions of categories and functors. The definition of persistence module relies on some common categories: The category $\mathbb{Z}^+$ (the category $\{1, \ldots, n\}$ alike) consists of objects from $\mathbb{Z}^+$ and a unique morphism from i to j if $i \leq j$. We also denote the morphism from i to j as $i \leq j$. The category Simp consists of objects which are all the simplicial complexes and morphisms which are simplicial maps. The category Vect consists of objects which are all the vector spaces over $\mathbb{Z}_2$ and morphisms which are linear maps. A persistence module $\mathcal{P}$ is then defined as a functor $\mathcal{P} : \mathbb{Z}^+ \to \mathsf{Vect}^1$.

A persistence module is usually induced by a filtration $\mathcal{F} = \mathcal{F}(K)$ of a simplicial complex K, where the filtration $\mathcal{F} : \varnothing = K_0 \subseteq K_1 \subseteq \ldots \subseteq K_m = K$ is a filtered sequence of subcomplexes of K such that K_{i+1} and K_i differ by one simplex σ_{i+1}. We can also interpret a filtration $\mathcal{F}$ as a functor $\mathcal{F} : \mathbb{Z}^+ \to \mathsf{Simp}$, where $\mathcal{F}(i) = K_i$ for $i \leq m$, $\mathcal{F}(i) = K$ for $i > m$, and a morphism $\mathcal{F}[i \leq j] : \mathcal{F}(i) \to \mathcal{F}(j)$ is the inclusion. Denoting $\mathsf{H}_q : \mathsf{Simp} \to \mathsf{Vect}$ as the qth homology functor with $\mathbb{Z}_2$ coefficients, the H_q-persistence module $\mathcal{P}_q^{\mathcal{F}}$ of $\mathcal{F}$ is obtained by composing the two functors H_q and $\mathcal{F}$, that is, $\mathcal{P}_q^{\mathcal{F}} = \mathsf{H}_q\mathcal{F}$. Specifically, $\mathcal{P}_q^{\mathcal{F}}(i) = \mathsf{H}_q(K_i)$ for $i \leq m$, $\mathcal{P}_q^{\mathcal{F}}(i) = \mathsf{H}_q(K)$ for $i > m$, and the morphism $\mathcal{P}_q^{\mathcal{F}}[i \leq j] : \mathsf{H}_q(K_i) \to \mathsf{H}_q(K_j)^2$ is the linear map induced by the inclusion.

A special class of persistence modules is the interval modules. Given an interval $[b, d) \subset \mathbb{Z}^+$, an interval module $\mathcal{I}^{[b,d)}$ is defined as: $\mathcal{I}^{[b,d)}(i) = \mathbb{Z}_2$ for $i \in [b, d)$ and $\mathcal{I}^{[b,d)}(i) = 0$ otherwise; $\mathcal{I}^{[b,d)}[i \leq j]$ is the identity map for $i, j \in [b, d)$ and $\mathcal{I}^{[b,d)}[i \leq j]$ is the zero map otherwise. By quiver theory, a H_q-persistence module obtained from a finite complex K has a unique decomposition $\mathcal{P}_q^{\mathcal{F}} = \bigoplus_{j \in J} \mathcal{I}^{[b_j, d_j)}$ in terms of interval modules, where $J \subset \mathbb{Z}$ is a finite index set [7]. Let $\mathrm{D}(\mathcal{P}_q^{\mathcal{F}}) = \{[b_j, d_j) \mid j \in J\}$ denote the set of intervals of the interval modules which $\mathcal{P}_q^{\mathcal{F}}$ decomposes into. Observe that $\mathrm{D}(\mathcal{P}_q^{\mathcal{F}})$ is also called

---

[1] Sometimes we also call a functor $\mathcal{P} : \{1, \ldots, n\} \to \mathsf{Vect}$ as a persistence module.
[2] $K_j = K$ when $j > m$.

the *barcode* or *persistence diagram* in the literature [17]. Sometimes we will abuse the notation slightly to write $D_q(\mathcal{F})$, where the argument is the filtration instead of the module $\mathcal{P}_q^{\mathcal{F}}$ it generates.

3   Persistent Basis and Cycles

Definition 1 (Persistent Basis). *An indexed set of q-cycles $\{c_j \mid j \in J\}$ is called a persistent q-basis for a filtration $\mathcal{F}$ if $\mathcal{P}_q^{\mathcal{F}} = \bigoplus_{j \in J} \mathcal{I}^{[b_j, d_j)}$ and for each $j \in J$ and $b_j \leq k < d_j$, $\mathcal{I}^{[b_j, d_j)}(k) = \{0, [c_j]\}$.*

Definition 2 (Persistent Cycle). *For an interval $[b, d) \in D(\mathcal{P}_q^{\mathcal{F}})$, a q-cycle c is called a persistent q-cycle for the interval, if one of the following holds:*

- *$d \neq +\infty$, c is a cycle in K_b containing σ_b, and c is not a boundary in K_{d-1} but becomes a boundary in K_d;*
- *$d = +\infty$ and c is a cycle in K_b containing σ_b.*

Remark 1. Note that the definition of persistent cycles for finite intervals is identical to that of [23, 26].

The following theorem characterizes each cycle in a persistent basis:

Theorem 1. *An indexed set of q-cycles $\{c_j \mid j \in J\}$ is a persistent q-basis for a filtration $\mathcal{F}$ if and only if $\mathcal{P}_q^{\mathcal{F}} = \bigoplus_{j \in J} \mathcal{I}^{[b_j, d_j)}$ and c_j is a persistent q-cycle for every interval $[b_j, d_j) \in D(\mathcal{P}_q^{\mathcal{F}})$.*

Proof. Suppose $\{c_j \mid j \in J\}$ is an indexed set of q-cycles satisfying the above conditions. For each $j \in J$, we construct an interval module $\mathcal{I}_j$, such that $\mathcal{I}_j(i) = \{0, [c_j]\}$ for $b_j \leq i < d_j$ and $\mathcal{I}_j(i) = 0$ otherwise. We claim that $\mathcal{P}_q^{\mathcal{F}} = \bigoplus_{j \in J} \mathcal{I}_j$. We first prove that $\mathcal{P}_q^{\mathcal{F}}(i) = \bigoplus_{j \in J} \mathcal{I}_j(i)$ for each $i \in \mathbb{Z}^+$, by proving that $\{[c_j] \mid j \in J, i \in [b_j, d_j)\}$ forms a basis of $\mathcal{P}_q^{\mathcal{F}}(i)$. Using mathematical induction, since σ_1 is a vertex, this is trivially true. Suppose for $i - 1$ this is true. If σ_i is neither positive nor negative, i.e., $H_q(K_{i-1}) \approx H_q(K_i)$ by the isomorphism induced from the inclusion, this is also trivially true for i. If σ_i is positive, suppose the corresponding interval of σ_i is $[b_{j'}, d_{j'})$ (note that $b_{j'} = i$ and $d_{j'}$ could possibly be $+\infty$). Since $\{[c_j] \mid j \in J, i - 1 \in [b_j, d_j)\}$ are still independent in $\mathcal{P}_q^{\mathcal{F}}(i)$ and $[c_{j'}]$ is not in the span of them, then $\{[c_j] \mid j \in J, i - 1 \in [b_j, d_j)\} \cup [c_{j'}] = \{[c_j] \mid j \in J, i \in [b_j, d_j)\}$ are independent in $\mathcal{P}_q^{\mathcal{F}}(i)$. Since the cardinality of $\{[c_j] \mid j \in J, i \in [b_j, d_j)\}$ equals the dimension of $\mathcal{P}_q^{\mathcal{F}}(i)$, it must form a basis of $\mathcal{P}_q^{\mathcal{F}}(i)$. If σ_i is negative, then there must be a $[c_{j'}]$ for a $j' \in J$ such that $d_{j'} = i$. For any $[c] \in \mathcal{P}_q^{\mathcal{F}}(i) = H_q(K_i)$, $[c] = \sum_{j \in J'} [c_j]$, where $J' \subseteq \{j \in J \mid i - 1 \in [b_j, d_j)\}$. If $j' \in J'$, then $[c] = \sum_{j \in J' - \{j'\}} [c_j]$, because $[c_{j'}] = 0$ in $H_q(K_i)$. Then $\{[c_j] \mid j \in J, i - 1 \in [b_j, d_j)\} - \{c_{j'}\} = \{[c_j] \mid j \in J, i \in [b_j, d_j)\}$ spans $H_q(K_i)$. This means that it also forms a basis of $H_q(K_i)$. It is then obvious that the direct sums of the maps of the interval modules are actually the maps of $\mathcal{P}_q^{\mathcal{F}}$, so $\{c_j \mid j \in J\}$ is a persistent q-basis for $\mathcal{F}$.

Suppose $\{c_j \mid j \in J\}$ is a persistent q-basis for $\mathcal{F}$. For each $j \in J$, c_j must not be in $K_{b_j - 1}$, because otherwise $[c_j]$ would be in the image of $\mathcal{P}_q^{\mathcal{F}}[b_j - 1 \leq b_j]$. It is obvious that c_j must contain σ_j. Note that for each $j \in J$ and each $i \in [b_j, d_j)$, $\mathcal{P}_q^{\mathcal{F}}[i \leq i + 1]([c_j]) = \mathcal{I}^{[b_j, d_j)}[i \leq i + 1]([c_j])$. Then for each $j \in J$ such that $d_j \neq +\infty$, $[c_j] \neq 0$ in $K_{d_j - 1}$ and $[c_j] = 0$ in K_{d_j}.

With Definition 2 and Theorem 1, it is true that for a persistent q-cycle c of an interval $[b, d) \in D_q(\mathcal{F})$, we can always form an interval module decomposition of $\mathcal{P}_q^{\mathcal{F}}$, where c is a representative cycle for the interval module of $[b, d)$.

4   Minimal Persistent q-Basis and Their Limitations

We have already defined persistent basis, the optimal versions of which are of particular interest because they capture more geometric information of the space. The cycles for an optimal (minimal) persistent basis have already been defined and studied in [20,23]. In particular, the author of [23] proposed an integer program to compute these cycles. Although these integer programs can be solved exactly by linear programs for certain cases [12], the integer program is NP-hard in general. This of course does not settle the question of whether the problem of computing minimal persistent 1-cycles is NP-hard or not. We prove that it is indeed NP-hard and thus has no hope of admitting a polynomial time algorithm unless P = NP.

Consider a simplicial complex K with each edge being assigned a non-negative weight. We refer to such K as a weighted complex. For a 1-cycle c in K, define its weight to be the sum of all weights of its edges.

Definition 3 (Minimal Persistent 1-Cycle and 1-Basis). *Given a filtration $\mathcal{F}$ on a weighted complex K, a minimal persistent 1-cycle for an interval of $D_1(\mathcal{F})$ is defined to be a persistent 1-cycle for the interval with the minimal weight. An indexed set of 1-cycles $\{c_j \mid j \in J\}$ is a minimal persistent 1-basis for $\mathcal{F}$ if for every $[b_j, d_j) \in D_1(\mathcal{F})$, c_j is a minimal persistent 1-cycle for $[b_j, d_j)$.*

We prove that the following special version of the problem of finding a minimal persistent 1-cycle is NP-hard. This special version reduces to the general version straightforwardly in polynomial time by assigning every edge a weight of 1.

Problem 1 (LST-PERS-CYC). Given a filtration $\mathcal{F} : \varnothing = K_0 \subseteq K_1 \subseteq \ldots \subseteq K_m = K$, and a finite interval $[b, d) \in D_1(\mathcal{F})$, find a 1-cycle with the least number of edges which is born in K_b and becomes a boundary in K_d.

Similar to [8], we reduce the NP-hard MAX-2SAT [24] problem to LST-PERS-CYC. MAX-2SAT is defined as:

Problem 2 (MAX-2SAT). Given N variables $x_1, \ldots, x_N$ and M clauses $c_1, \ldots, c_M$, with the clauses being the disjunction of at most two variables. Find an assignment of Boolean values to all the variables such that the maximal number of clauses are satisfied.

Theorem 2. *The problem LST-PERS-CYC is NP-hard*

Proof (sketch). Because of the space restriction, we only give a sketch of the reduction from MAX-2SAT to LST-PERS-CYC. The entire proof can be seen in the full version of this paper. Given an instance of MAX-2SAT, we first construct a simplicial complex K as in [8], by forming a triangulated cylinder C_i for each variable x_i and a cycle w_j for each clause c_j, such that the two ends z_i and z_i' of C_i correspond to x_i and $\neg x_i$. Each z_i or z_i' has $3M$ edges and each w_j has three edges. We make z_i's and z_i'''s share edges with w_j's such that there is an assignment of Boolean values making k clauses satisfied if and only if there is a cycle in $[\overline{z}]$ with $3MN + 3M - 2k$ edges, where $\overline{z} = \sum_{i=1}^{N} z_i + \sum_{j=1}^{M} w_j$.

We construct a filtration $\mathcal{F}$ of K, with the restriction: Pick an edge e of a clause cycle, which is not shared with any end cycle of the variable cylinders, and take e as the last simplex added to the filtration. We then construct a filtration $\mathcal{F}'$ appending some simplices to $\mathcal{F}$ such that the only negative triangle of $\mathcal{F}' \setminus \mathcal{F}$ is the last added triangle t, and the adding of t only make $[\overline{z}]$ become 0. Let the indices of e and t in $\mathcal{F}'$ be b and d, then $[b, d)$ is an interval of $\mathrm{D}_1(\mathcal{F}')$. Now we have an instance of LST-PERS-CYC: Given the filtration $\mathcal{F}'$ and the interval $[b, d) \in \mathrm{D}_1(\mathcal{F})$, find a persistent 1-cycle with the least number of edges. Computing the answer to the instance of LST-PERS-CYC gives the answer to the instance of MAX-2SAT. Furthermore, the reduction is in polynomial time and the size of the constructed instance of LST-PERS-CYC is a polynomial function of that of MAX-2SAT, so LST-PERS-CYC is NP-hard.

4.1 Instability of Minimal Persistent 1-Cycles

In this section, we inspect if the minimal persistent 1-cycles are stable against perturbation of the filtration. Note that there may be multiple minimal persistent 1-cycles for an interval and an algorithm may choose anyone of them. This means that the cycles cannot be stable under those measures that take into account the entire geometry of the cycles (e.g., Hausdorff distance). In an attempt to sidestep this problem, we take a 'weaker' measure of the cycles which is still meaningful, namely their lengths. We show that even under such a measure, minimal persistent 1-cycles are unstable. Specifically, we consider the lower star filtration [17] of a vertex sequence, and inspect the perturbation of the lengths of persistent 1-cycles under the perturbation of the sequence. Since each interval I in the H_1-persistence diagram of a lower star filtration can be derived from an interval I' in the H_1-persistence diagram of a corresponding insertion filtration[3], we can associate a persistent 1-cycle for I' to I. The readers can verify that this assignment gives representatives for the decomposed interval modules of the H_1-persistence module induced by the lower star filtration.

Figure 2a presents an example for which the perturbation of the minimal persistent 1-cycles cannot be properly bounded. The object in Fig. 2a is a sphere with two holes (i.e., c_1 and c_2). We can assume that the object is nicely triangulated so that it becomes a simplicial complex. Let v_1 and v_2 be vertices from c_1

---

[3] The *insertion filtration* is actually the filtration defined in Sect. 2.

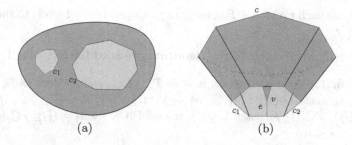

(a)   (b)

Fig. 2. (a) A sphere with two holes shows the instability of the minimal persistent 1-cycles. (b) The simplicial complex shows the instability of the cycles of Algorithm 2. c_1 and c_2 are the two red simple cycles; e is the edge adjacent to v. (Color figure online)

and c_2. We can construct a filtration[4] by first forming the two cycles c_1 and c_2, with v_1 and v_2 being the last two vertices added, then adding the other parts of the simplicial complex. We then add a cone around c_1 to the filtration. We can first assume v_1 is added before v_2, and the indices of v_1 and the apex vertex of the cone in the sequence are b and d. Then the minimal persistent 1-cycle for the interval $[b, d)$ is c_1. If we switch v_1 and v_2, the minimal cycle for the interval $[b, d)$ becomes c_2. The difference of c_1 and c_2 can be made arbitrary under a single switch, which is the smallest possible perturbation of lower star filtration.

5   Computing Meaningful Persistent 1-Cycles in Polynomial Time

Because the minimal persistent 1-cycles are not stable and their computation is NP-hard, we propose an alternative set of meaningful persistent 1-cycles which can be computed efficiently in polynomial time. We first present a general framework. Although the computed persistent 1-cycles have no guaranteed properties, the framework lays the foundation for the algorithm computing meaningful persistent 1-cycles that we propose later.

Algorithm 1. Given a simplicial complex K, a filtration $\mathcal{F} : \varnothing = K_0 \subseteq K_1 \subseteq \ldots \subseteq K_m = K$, and $D_1(\mathcal{F})$, this algorithm finds a persistent 1-basis for $\mathcal{F}$. The algorithm maintains a basis $\mathcal{B}_i$ for $H_1(K_i)$ for every $i \in [0, m]$. Initially, let $\mathcal{B}_0 = \varnothing$, then do the following for $i = 1, \ldots, m$:

- If σ_i is positive, find a 1-cycle c_i containing σ_i in K_i and let $\mathcal{B}_i = \mathcal{B}_{i-1} \cup \{c_i\}$.
- If σ_i is negative, find a set $\{c_g \mid g \in G\} \subseteq \mathcal{B}_{i-1}$ so that $\sum_{g \in G} [c_g] = 0$. This can be done in $O(\beta_i = |\mathcal{B}_i|)$ time by the annotation algorithm in [13]. Maintaining the annotations will take $O(n^\omega)$ time altogether where K has n simplices and ω is the matrix multiplication exponent. Let g^* be the greatest index in G, then $[g^*, i)$ is an interval of $D_1(\mathcal{F})$. Assign $\sum_{g \in G} c_g$ to this interval as a persistent 1-cycle and let $\mathcal{B}_i = \mathcal{B}_{i-1} \smallsetminus c_{g^*}$.
- Otherwise, let $\mathcal{B}_i = \mathcal{B}_{i-1}$.

---

[4] Note that we are constructing an insertion filtration for a lower star filtration.

At the end, for each cycle $c_j \in \mathcal{B}_m$, assign c_j as a persistent 1-cycle to the interval $[j, +\infty)$.

To prove the correctness of the algorithm, we need the following fact:

Proposition 1. *For a persistence module* $\mathcal{P} : \{1, \ldots, n\} \to \mathsf{Vect}$ *and a finite set of persistence modules* $\{\mathcal{Q}_j : \{1, \ldots, n\} \to \mathsf{Vect} | j \in J\}$, $\mathcal{P} = \bigoplus_{j \in J} \mathcal{Q}_j$ *if and only if* $\mathcal{P}(i) = \bigoplus_{j \in J} \mathcal{Q}_j(i)$ *for each* $1 \le i \le n$ *and* $\mathcal{P}[i \le i+1] = \bigoplus_{j \in J} \mathcal{Q}_j[i \le i+1]$ *for each* $1 \le i < n$.

Proof (Correctness of Algorithm 1). Denoting all the intervals $[g^*, i)$ found by the algorithm as D, we want to inductively prove that for all $i = 1, \ldots, m$, the persistence module $\mathcal{P}^i$, which is the restriction of $\mathcal{P}_1^{\mathcal{F}}$ to $\{1, ..., i\}$, satisfies:

$$\mathcal{P}^i = \bigoplus_{[b_j, d_j) \in D, d_j \le i} \mathcal{I}^{[b_j, d_j)} \oplus \bigoplus_{c_j \in \mathcal{B}_i} \mathcal{I}^{[j, i]} \tag{1}$$

where the representative of $\mathcal{I}^{[b_j, d_j)}$ is the persistent 1-cycle computed by the algorithm and the representative of $\mathcal{I}^{[j, i]}$ is c_j. When $i = 1$, $\mathcal{P}^1$ is trivial and the equation is certainly true. Suppose for $\mathcal{P}^i$, the equation is satisfied. If σ_{i+1} is neither positive nor negative, or positive, then it is not hard to verify that the equation is still valid for $\mathcal{P}^{i+1}$ by Proposition 1. If σ_{i+1} is negative, then we can let the persistent 1-cycle computed by the algorithm for σ_{i+1} be $\sum_{g \in G} c_g$ and g^* be the greatest index in G. Since $\sum_{g \in G} c_g$ is also created by σ_{g^*}, we can let the representative of the interval module $\mathcal{I}^{[g^*, i]}$ for $\mathcal{P}^i$ be $\sum_{g \in G} c_g$. It is not hard then to verify that the equation is still satisfied for $\mathcal{P}^{i+1}$ by Proposition 1.

Based on Algorithm 1, we present another algorithm which produces meaningful persistent 1-cycles.

Algorithm 2. In Algorithm 1, when σ_i is positive, let c_i be the shortest cycle containing σ_i in K_i. The cycle c_i can be constructed by adding σ_i to the shortest path between vertices of σ_i in K_{i-1}, which can be computed by Dijkstra's algorithm applied to the 1-skeleton of K_{i-1}.

Note that in Algorithm 2, a persistent 1-cycle for a finite interval is a sum of shortest cycles born at different indices. Since a shortest cycle is usually a good representative of its class, the sum of shortest cycles ought to be a good choice of representative for an interval. In some cases, when σ_i is negative, the sum $\sum_{g \in G} c_g$ contains only one component. The persistent 1-cycles computed by Algorithm 2 for such intervals are guaranteed to be optimal as shown below.

Proposition 2. *In Algorithm 2, when* σ_i *is negative, if* $|G| = 1$, *then* $\sum_{g \in G} c_g$ *is a minimal persistent 1-cycle for the interval ending with* i.

In Sect. 6 where we present the experimental results, we can see that, scenarios depicted by Proposition 2 occur quite frequently. Especially, for the larvae and nerve datasets, nearly all computed persistent 1-cycles contain only one component and hence are minimal.

A practical problem with Algorithm 2 is that unnecessary computational resource is spent for computing tiny intervals regarded as noise, especially when the user cares about significantly large intervals only. We present a more efficient algorithm for such cases.

Proposition 3. *In Algorithms 1 and 2, when σ_i is negative, for any $g \in G$, one has $b_g \leq g^*$ and $d_g \geq i$.*

Proof. Note that σ_{b_g} must be unpaired before σ_i is added, this implies that $d_g \geq i$. Since g^* is the greatest index in G, $b_g = g \leq g^*$.

Proposition 3 leads to Algorithm 3 in which we only compute the shortest cycles at the birth indices whose corresponding intervals contain the input interval $[b, d)$. In the worst case, Algorithms 2 and 3 run in $O(n^\omega + n^2 \log n) = O(n^\omega)$ time. However, since an user often provides a long interval, the intervals containing it constitute a small subset of all the intervals. This makes Algorithm 3 run much faster than Algorithm 2 in practice.

Algorithm 3

Input: The input of Algorithm 2 plus an interval $[b, d)$
Output: A persistent 1-cycle for $[b, d)$ output by Algorithm 2.
1: $G' \leftarrow \varnothing$
2: **for** $i \leftarrow 1, \ldots, b$ **do**
3:     **if** σ_i is positive **and** (σ_i is paired with a σ_j s.t $j \geq d$
        or σ_i never gets paired) **then**
4:         $c_i \leftarrow$ the shortest cycle containing σ_i in K_i
5:         $G' \leftarrow G' \cup \{i\}$
6:     **end if**
7: **end for**
8: find a $G \subseteq G'$ s.t. $\sum_{g \in G}[c_g] = 0$ in K_d
9: output $\sum_{g \in G} c_g$ as the persistent 1-cycle for $[b, d)$

Proposition 4 reveals some characteristics of the persistent 1-cycles computed by Algorithms 2 and 3:

Proposition 4 (Minimality Property). *The persistent 1-cycle $\sum_{g \in G} c_g$ computed by Algorithms 2 and 3 has the following property: There is no non-empty proper subset G' of G such that $\sum_{g \in G'}[c_g] = 0$ in $H_1(K_d)$, where d is the death index of the interval to which $\sum_{g \in G} c_g$ is associated.*

Given that the minimal persistent 1-cycles are not stable, it is not surprising that the cycles computed by Algorithm 2 are also not stable under perturbation.

Figure 2b presents an example for which the perturbation of persistent 1-cycles computed by Algorithm 2 cannot be properly bounded. We can construct a filtration by first forming the cycle c then adding the other parts of the simplicial complex in Fig. 2b, making v the last vertex and e the last simplex. We then add a cone around c_1 to the filtration. Let the indices of v and the apex vertex of the cone in the vertex sequence be b and d. When c is formed, the last edge e' of c is positive, and c is chosen as the shortest cycle containing e'. When e is added, we can make c_1 and c_2 be the two shortest cycles containing e. When c_1 is coned, if c_1 is chosen as the shortest cycle containing e, then the persistent 1-cycle for the interval $[b, d)$ would be c_1. Otherwise, the persistent 1-cycle would be $c + c_2$. The length of c can be arbitrary, so that the difference of the two persistent 1-cycles can be arbitrary under the same insertion filtration of the same lower star filtration.

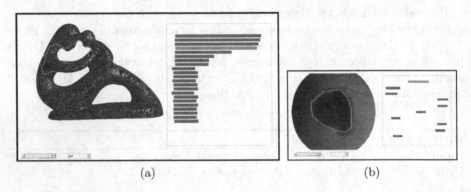

(a)   (b)

Fig. 3. PersLoop user interface demonstrating persistent 1-cycles computed for a 3D point cloud (a) and a 2D image (b), where green cycles correspond to the chosen bars. (Color figure online)

6   Results and Experiments

Our software PersLoop implements Algorithm 3. Given a raw input, which is a 2D image or a 3D point cloud, and a filtration built from the raw input, the software first generates and plots the barcode of the filtration. The user can then click an individual bar to obtain the persistent 1-cycle for the bar (Fig. 3). The experiments on 3D point clouds and 2D images using the software show how our algorithm can find meaningful persistent 1-cycles in several geometric and vision related applications.

6.1   Persistent 1-Cycles for 3D Point Clouds

We take a 3D point cloud as input and build a Rips filtration using the Gudhi library [25]. As shown in Fig. 4, persistent 1-cycles computed for the three point

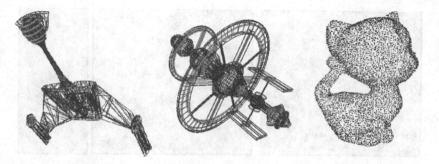

Fig. 4. Persistent 1-cycles (green) corresponding to long intervals computed for three different point clouds (Color figure online)

clouds sampled from various models are tight and capture essential geometrical features of the corresponding persistent homology. Note that our implementation of Algorithm 3 runs very fast in practice. For example, it took 0.3 secs to generate 50 cycles on a regular commodity laptop for the Botijo (Fig. 1a) point cloud of size 10,000.

6.2 Image Segmentation and Characterization Using Cubical Complex

In this section, we show the application of our algorithm on image segmentation and characterization problems. We interpret an image as a piecewise linear function on a 2-dimensional cubical complex. The cubical complex for an image has a vertex for each pixel, an edge connecting each pair of horizontally or vertically adjacent vertices, and squares to fill all the holes such that the complex becomes a disc. The function values on the vertices are the density or color values of the corresponding pixels. The lower star filtration [17] of the PL function is then built and fed into our software. We use the coning based annotation strategy [13] to compute the persistence diagrams. In our implementation, a cubical tree, which is similar to the simplicial tree [3], is built to store the elementary cubes. Each elementary cube points to a row in the annotation matrix via the union find data structure. The simplicial counterpart of this association technique is described in [2].

Our first experiment is the segmentation of a serial section Transmission Electron Microscopy (ssTEM) data set of the Drosophila first instar larva ventral nerve cord (VNC) [5]. The segmentation result is shown in Fig. 5a and b, from which we can see that the cycles are in exact correspondence to the membranes hence segment the nerve regions quite appropriately. The difference between Fig. 5a and b shows that longer intervals tend to have longer cycles. Another similar application is the segmentation of the low magnification micrographs of a Drosophila embryo [22]. As seen in Fig. 5c, the cycles corresponding to the top 200 longest intervals indicate that the larvae image is properly segmented.

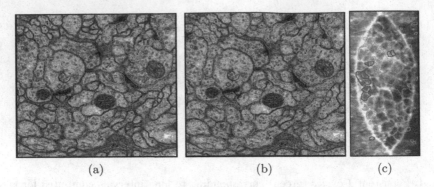

(a)   (b)   (c)

Fig. 5. Persistent 1-cycles computed for image segmentation. Green cycles indicate persistent 1-cycles consisting of only one component ($|G| = 1$) and red cycles indicate those consisting of multiple components ($|G| > 1$). (a,b) Persistent 1-cycles for the top 20 and 350 longest intervals on the nerve dataset. (c) Persistent 1-cycles for the top 200 longest intervals on the Drosophila larvae dataset. (Color figure online)

We experiment on another dataset from the STARE project [21] to show how persistent 1-cycles computed by our algorithm can be utilized for characterization of images. The dataset contains ophthalmologist annotated retinal images which are either healthy or suffering from diseases. Our aim is to automatically identify retinal and sub-retinal hemorrhages, which are black patches of blood accumulated inside eyes. Figures 1e and 3b show that a very tight cycle is derived around each dark hemorrhage blob even when the input is noisy.

6.3   Hexagonal Structure of Crystalline Solids

In this experiment, we use our persistent 1-cycles to describe the crystalline structure of silicate glass (SiO_2) commonly known as quartz. Silicate glass has a non-compact structure with three silicon and oxygen atoms arranged alternately in a hexagon as shown in Fig. 6a. We build a $8 \times 8 \times 8$ weighted point cloud with the silicon and oxygen atoms arranged according to the space group on the crystal structure as illustrated in Fig. 6b. The weights of the points correspond to the atomic weights of the atoms. On this weighted point cloud, we generate a filtration of weighted alpha complexes [16] by increasing α from 0 to ∞.

Persistent 1-cycles computed by our algorithm for this dataset reveal both the local and global structures of silicate glass. Figure 6d lists the barcode of the filtration we build and Fig. 6b shows the persistent 1-cycles corresponding to the medium sized green bars in Fig. 6d. We can see on close observation that the cycles in Fig. 6b are in exact accordance to the hexagonal cyclic structure of quartz shown in Fig. 6a. The larger persistent 1-cycles in Fig. 6c, which span the larger lattice structure formed by our weighted point cloud, correspond to the longer red bars in Fig. 6d. These cycles arise from the *long-range order*[5] of

---

[5] Long-range order is the translational periodicity where the self-repeating structure extends infinitely in all directions.

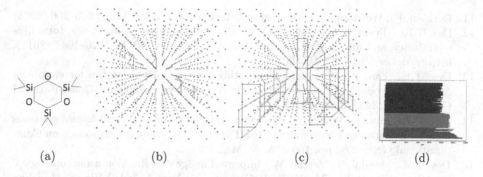

(a)   (b)   (c)   (d)

Fig. 6. (a) Hexagonal cyclic structure of silicate glass. (b) Persistent 1-cycles computed for the green bars with red points denoting silicate atoms and grey points denoting oxygen atoms. (c) Persistent 1-cycles computed for the red bars. (d) Barcode for the filtration on silicate glass. (Color figure online)

the crystalline solid. This is evident from our experiment because if we increase the size of the input point cloud, these cycles grow larger to span the entire lattice.

References

1. Awodey, S.: Category Theory. Oxford University Press, Oxford (2010)
2. Boissonnat, J., Dey, T.K., Maria, C.: The compressed annotation matrix: an efficient data structure for computing persistent cohomology. CoRR abs/1304.6813 (2013). http://arxiv.org/abs/1304.6813
3. Boissonnat, J.D., Maria., C.: The simplex tree: an efficient data structure for general simplicial complexes. In: 20th Annual European Symposium, Ljubljana, Slovenia, vol. 2, pp. 731–742 (2012)
4. Bubenik, P., Scott, J.A.: Categorification of persistent homology. Discret. Comput. Geom. **51**(3), 600–627 (2014)
5. Cardona, A., Saalfeld, S., Preibisch, S., Schmid, B., Cheng, A., Pulokas, J., et al.: An integrated micro- and macroarchitectural analysis of the drosophila brain by computer-assisted serial section electron microscopy. PLoS Biol. **8**, e1000502 (2010)
6. Carlsson, G., Ishkhanov, T., de Silva, V., Zomorodian, A.: On the local behavior of spaces of natural images. Int. J. Comput. Vis. **76**(1), 1–12 (2008). https://doi.org/10.1007/s11263-007-0056-x
7. Chazal, F., de Silva, V., Glisse, M., Oudot, S.: The Structure and Stability of Persistence Modules. Springer, Cham (2016). https://doi.org/10.1007/978-3-319-42545-0
8. Chen, C., Freedman, D.: Quantifying homology classes II: localization and stability. arXiv preprint arXiv:0709.2512 (2007)
9. Chen, C., Freedman, D.: Hardness results for homology localization. Discret. Comput. Geom. **45**(3), 425–448 (2011)
10. Cohen-Steiner, D., Edelsbrunner, H., Harer, J.: Stability of persistence diagrams. In: Proceedings of the Twenty-First Annual Symposium on Computational Geometry, pp. 263–271. ACM (2005)

11. Derksen, H., Weyman, J.: Quiver representations. Not. AMS **52**(2), 200–206 (2005)
12. Dey, T.K., Hirani, A., Krishnamoorthy, B.: Optimal homologous cycles, total uni-modularity, and linear programming. SIAM J. Comput. **40**(4), 1026–1044 (2011). https://doi.org/10.1137/100800245
13. Dey, T.K., Fan, F., Wang, Y.: Computing topological persistence for simplicial maps. In: Proceedings of the Thirtieth Annual Symposium on Computational Geometry, p. 345. ACM (2014)
14. Dey, T.K., Sun, J., Wang, Y.: Approximating loops in a shortest homology basis from point data. In: Proceedings of the Twenty-Sixth Annual Symposium on Computational Geometry, pp. 166–175. ACM (2010)
15. Dey, T.K., Mandal, S., Varcho, W.: Improved image classification using topological persistence. In: Hullin, M., Klein, R., Schultz, T., Yao, A. (eds.) Vision, Modeling & Visualization. The Eurographics Association (2017). https://doi.org/10.2312/vmv.20171272
16. Edelsbrunner, H.: Weighted alpha shapes. Technical report, Champaign, IL, USA (1992)
17. Edelsbrunner, H., Harer, J.: Computational Topology: An Introduction. American Mathematical Society, Providence (2010)
18. Edelsbrunner, H., Letscher, D., Zomorodian, A.: Topological persistence and sim-plification. In: Proceedings of the 41st Annual Symposium on Foundations of Computer Science, pp. 454–463. IEEE (2000)
19. Emmett, K., Schweinhart, B., Rabadan, R.: Multiscale topology of chromatin fold-ing. In: Proceedings of the 9th EAI International Conference on Bio-inspired Information and Communications Technologies (formerly BIONETICS), pp. 177–180. ICST (Institute for Computer Sciences, Social-Informatics and Telecommunications Engineering) (2016)
20. Escolar, E.G., Hiraoka, Y.: Optimal cycles for persistent homology via linear programming. In: Fujisawa, K., Shinano, Y., Waki, H. (eds.) Optimization in the Real World, vol. 13, pp. 79–96. Springer, Tokyo (2016). https://doi.org/10.1007/978-4-431-55420-2_5
21. Hoover, A., Goldbaum, M.: Locating the optic nerve in a retinal image using the fuzzy convergence of the blood vessels. IEEE Trans. Med. Imaging **22**(8), 951–958 (2003). https://doi.org/10.1109/TMI.2003.815900
22. Kiehart, D.P., Galbraith, C.G., Edwards, K.A., Rickoll, W.L., Montague, R.A.: Multiple forces contribute to cell sheet morphogenesis for dorsal closure in Drosophila. J. Cell Biol. **149**(2), 471–490 (2000). https://doi.org/10.1083/jcb.149.2.471. http://jcb.rupress.org/content/149/2/471
23. Obayashi, I.: Volume optimal cycle: tightest representative cycle of a generator on persistent homology. arXiv preprint arXiv:1712.05103 (2017)
24. Papadimitriou, C.H., Yannakakis, M.: Optimization, approximation, and complexity classes. J. Comput. Syst. Sci. **43**(3), 425–440 (1991)
25. The GUDHI Project: GUDHI User and Reference Manual. GUDHI Editorial Board (2015). http://gudhi.gforge.inria.fr/doc/latest/
26. Wu, P., et al.: Optimal topological cycles and their application in cardiac trabeculae restoration. In: Niethammer, M., et al. (eds.) IPMI 2017. LNCS, vol. 10265, pp. 80–92. Springer, Cham (2017). https://doi.org/10.1007/978-3-319-59050-9_7

A Persistence-Based Approach to Automatic Detection of Line Segments in Images

Vitaliy Kurlin[1]([✉]) [iD] and Grzegorz Muszynski[1,2]

[1] Department of Computer Science, University of Liverpool, Liverpool L69 3BX, UK
vkurlin@liverpool.ac.uk
[2] Lawrence Berkeley National Laboratory, Berkeley, CA, USA
gmuszynski@lbl.gov

Abstract. Edge detection algorithms usually produce a discrete set of edgels (edge pixels) in a given image on a fixed pixel grid. We consider the harder problem of detecting continuous straight line segments at subpixel resolution. The state-of-the art Line Segment Detection Algorithm (LSDA) outputs unordered line segments whose total number cannot be easily controlled. Another motivation to improve the LSDA is to avoid intersections and small angles between line segments, hence difficulties in higher level tasks such as segmentation or contour extraction.

The new Persistent Line Segment Detector (PLSD) outputs only non-intersecting line segments and ranks them by a strength, hence the user can choose a number of segments. The main novelty is an automatic selection of strongest segments along any straight line by using the persistence from Topological Data Analysis. The experiments on the Berkeley Segmentation Database of 500 real-life images show that the new algorithm outperforms the LSDA on the important measure of Boundary Recall.

Keywords: Topological persistence · Edge detection · Skeletonization

1 Introduction

1.1 The Edge Detection Problem in the Continuous Setting

Detecting edges in images is a key problem in the low-level vision that aims to identify pixels where the image intensity suddenly changes. The edge detection was usually considered in the discrete setting when an output consists of pixels from a given pixel grid. However, pixel-based images represent a continuous world, where the most basic objects are continuous line segments, which may have arbitrary directions and endpoints with any real coordinates.

The hard version of edge detection is to find straight line segments at subpixel resolution that approximate boundary contours in pixel-based images.

The state-of-the-art algorithm [11] solving the above problem is the Line Segment Detection Algorithm (LSDA). The main advantage of the LSDA over past edge detection algorithms is the "a contrario" approach that theoretically guarantees at most one false alarm on random data, see details in Subsect. 2.2.

© Springer Nature Switzerland AG 2019
R. Marfil et al. (Eds.): CTIC 2019, LNCS 11382, pp. 137–150, 2019.
https://doi.org/10.1007/978-3-030-10828-1_11

1.2   Motivations to Detect Line Segments Without Intersections

The LSDA often outputs line segments that intersect each other near their endpoints, see 48 intersections for the image in Fig. 1 from the Berkeley Segmentation Database [1]. Duan and Lafarge [6] have proposed a refinement of the LSDA edge for producing Voronoi superpixels at subpixel resolution. This refinement has revealed that some LSDA edges are too close and almost parallel to each other as clearly illustrated in [10, Fig. 1.1 on p. 1]. So these close lines should be removed or carefully repaired to avoid very narrow superpixels. Figure 1 shows how the PLSD avoids all intersections of edges in comparision with the LSDA.

Fig. 1. Left: 193 LSDA edges with 48 intersections. **Right**: the PLSD outputs exactly 100 edges without any intersections of edges, which is the key advantage over LSDA.

The hard difficulties above are understandable taking into account that the LSDA attempts to capture line segments with any possible slope. Since approximate solutions are acceptable in real-life, we simplify the problem and will detect line segments that are parallel to one of 8 directions: horizontal $(1, 0)$, vertical $(0, 1)$, two diagonal $(\pm 1, 1)$ and four non-diagonal directions $(\pm 2, 1)$, $(\pm 1, 2)$.

We believe that 8 directions are enough to approximate any reasonable shapes in images, e.g. a large round disk in Fig. 2 can be well approximated by polygonal curves with 16 edges split into 8 pairs of opposite parallel edges.

One more important motivation to improve the LSDA is to control the number of edges in a final output. When LSDA edges are included into a polygonal mesh, the size of the mesh (number of polygons) may depend on the number of original edges. Hence, it would be great to order detected edges by some sort of strength so that a smaller number of strongest edges can be selected.

1.3   Automatic Selection of Persistent Segments

The main novelty of the proposed algorithm PLSD (Persistent Line Segment Detector) is the automatic selection of strongest segments in any straight line.

A grayscale image on $\Omega = [0, w] \times [0, h]$ is a function $I : \Omega \to [0, 255]$ sampled at pixel positions $p \in \Omega$ with integer coordinates in the image $[0, w] \times [0, h]$. An edge detection algorithm outputs pixels $p_1, \ldots, p_k \in \Omega$, where the function I substantially changes (depending on an algorithm) along some direction. This change at a fixed pixel is measured as the magnitude of the image gradient.

For a function $f : L \to \mathbb{R}$ of contrast values along a fixed straight line L in an image, we analyze the sequence of superlevel sets $f^{-1}[u, +\infty) = \{p \in L : f(p) \geq u\}$. For every fixed level u of the contrast, the superlevel set splits into a few continuous segments over which the contrast is at least u.

When the contrast level u goes down, new segments appear around local maxima of f and then merge with each other, see Fig. 4. So each segment S *persists* from its *birth* (at the maximum value of u) to its *death* (at the value when S merges with another segment having a higher birth), see Definition 2.

A segment S is usually characterized by its persistence = birth-death (when the parameter u is decreasing). We suggest another characteristic (the *strength* $|S| = \int_S f(p) dp$), which is more stable under perturbations of contrast values, hence is more suitable for noisy data, see formal details in Definition 3.

Line segments are ranked according to the concept of persistence, which was introduced in Topological Data Analysis [7]. The idea of persistence is to study a nested sequence of shapes parameterized across all potential thresholds.

At every level u the strongest segments are separated from noisy artefacts by a widest gap in strength, which is the maximum difference between successive ordered strength values over all current segments, see Definition 3.

The same widest gap in persistence was successfully used for segmenting clouds of points [14–17] that are not restricted to a fixed pixel grid as in digital images. So the strongest segments are independently selected along every straight line L considered in an image. Hence there is no uniform thresholding for the whole image, see details of this new automatic method in Subsect. 3.2.

Here is the summary of key contributions.

- The edge detection is studied in the continuous setting, which is harder than for discrete square-based pixels.
- The algorithm PLSD can output a desired number of strongest straight line segments that have no intersections guaranteed by Stage 2 in Subsect. 3.3.
- The main innovation of the Persistent Line Segment Detector is a data-driven automatic selection of persistent line segments without manual thresholding.
- The PLSD runs in a near linear time, see Theorem 5, and outperforms the state-of-the-art Line Segment Detector on the Boundary Recall benchmark from the Berkeley Segmentation Database 500 [1].

2 Review of the Past Closely Related Work

This section discusses a few representative algorithms for detecting only straight line segments at subpixel precision.

Fig. 2. The new algorithm PLSD outputs line segments in 8 directions, which can well approximate complicated shapes, even a large round disk in the last image above.

2.1 From Discrete Pixels to Continuous Arcs

Many past algorithms are based on the famous Canny detector of edge pixels [3], which already requires three parameters. The next usual step is to apply a Hough transform [2] to find lines passing through a certain number of edgels.

The Hough transform often leads to many false positives in textured regions. Another approach by Kahn et al. [12] uses only orientations of image gradients, but not their magnitudes. Their algorithm produces well localized edges, but requires carefully chosen thresholds.

A different "a contrario" (by contraries) approach is to validate potential candidates by setting thresholds on random data as follows. If a parametric algorithm on random data outputs a small number of false positives on average, the corresponding thresholds should be fixed and applied to real data. The only drawback was the exhaustive search through $O(P^4)$ possible straight lines, where P is the perimeter of an image. This method has led to the fast LSDA below.

2.2 The State-of-the-Art Line Segment Detection Algorithm (LSDA)

The LSDA outputs line segments detected in a grayscale image at subpixel resolution [11]. The first step is to estimate the image gradient dI as the vector (g_x, g_y) whose components are obtained by convolving with these 2×2 masks:

$$g_x = \begin{bmatrix} -1 & +1 \\ -1 & +1 \end{bmatrix}, \qquad g_y = \begin{bmatrix} +1 & +1 \\ -1 & -1 \end{bmatrix}. \qquad (2.2)$$

The operators above estimate the image derivatives in the x, y directions at the corner point shared by 4 pixels $(x, y), (x, y + 1), (x + 1, y), (x + 1, y + 1)$. So ideal edges were expected to be along boundaries of square pixels, but the original LSDA code shifted the final edges by $(0.5, 0.5)$. After normalising the gradient by its Euclidean length, the resulting field consists of unit length vectors.

Pixels whose estimated unit vectors are almost parallel (within a default tolerance $\tau = 22.5°$ for angles) are clustered. The resulting clusters are approximated by thin rectangles whose long middle lines are the final line segments. The output is an unordered list of line segments whose total number depends on a given image, so users may struggle to get a specific number of line segments.

2.3 Applications of Line Segments for Superpixels

Since rectangles covering adjacent clusters may overlap, LSDA edges may have intersections close to their endpoints. The LSDA outputs line segments with intersections on about 80% of 500 BSD images without any order. Hence any further application of the LSDA for segmentation or contour extraction requires a careful refinement of LSDA edges. The LSDA output was used for Voronoi superpixels by Duan and Lafarge [6], who designed a multi-step post-processing to repair segments that intersect each other or have very close endpoints.

The main result of Duan and Lafarge [6] is probably the first algorithm splitting an image into convex polygons whose vertices may have any real coordinates. Forsythe and Kurlin [8,9] used a more sophisticated refinement of the LSDA output and proved that the resulting Convex Constrained Meshes (CCM) have no small angles and approximate LSDA edges considered as hard constraints.

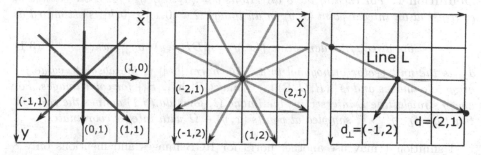

Fig. 3. Left: first 4 basic directions of line segments in the current implementation of the PLSD. **Middle**: for more directions. **Right**: the contrast function $f_L : L \to \mathbb{R}$ from Definition 1 is computed at all red points $(x, y) \in L$ with both integer coordinates. (Color figure online)

The new detector PLSD can be used in both methods above without extra refinement, because all final edges have no intersections by construction.

3 PLSD: The New Persistent Line Segment Detector

This section describes the following 3 stages of the PLSD algorithm.

Stage 1: estimating the change of contrast along every straight line $d_x x + d_y y + t = 0$, where (d_x, d_y) is one of the 8 slopes in Fig. 3, the shift t takes all integer values when the resulting line L intersects the image $\Omega = [0, w] \times [0, h]$.
Stage 2: automatic selection of strongest line segments by their persistence using the contrast function along every line $d_x x + d_y y + t = 0$ from Stage 1.
Stage 3: choosing a required number of strongest segments (one by one) so that any weaker segments don't intersect already chosen stronger segments.

3.1   Stage 1: Computing the Contrast Functions f Along Lines L

The first step convolves a given image I with the Gaussian kernel 3×3 with the default parameter $\sigma = 0.8$ using the GaussianBlur function in OpenCV. The second step considers all straight lines that intersect the image and are parallel to one of the 8 directions: $(1,0)$, $(0,1)$, $(\pm1,1)$, $(\pm1,2)$, $(\pm2,1)$ in Fig. 3. These 8 directions are chosen for simplicity and speed of the current implementation.

Let the image domain Ω be a rectangle $[0,w] \times [0,h]$. Then we consider all points (x,y) with integer coordinates $b \leq x \leq w - b$, $b \leq y \leq h - b$. Here b is a small offset (the default value 3 pixels) that allows us to convolve I with gradient masks and avoid boundary effects. For a fixed point (x,y) with integer coordinates, the current implementation uses the simplest 2×2 masks g_x, g_y in formulae (2.2) to estimate the image gradient as $DI = (g_x * I, g_y * I)$. If I is a color image, the same linear operators g_x, g_y are applied to every color channel.

Definition 1. *For each of the 8 directions $d = (d_x, d_y)$ in Fig. 3, the change of contrast at an integer point (x,y) in an image $\Omega = [0,w] \times [0,h]$ is estimated as*

$$\text{the directional derivative } f(x,y) = \|DI(x,y) \cdot d_\perp\|, \text{ where} \qquad (3.1)$$

$d_\perp$ is the unit vector orthogonal to d. The norm $\| \cdot \|$ is the absolute value for grayscale images and is $\|(R,G,B)\|_\infty = \max\{|R|, |G|, |B|\}$ for color images. For every straight line L intersecting the image Ω, formula (3.1) defines the contrast function $f_L : L \to \mathbb{R}$ sampled at points $(x,y) \in \Omega$ with integer coordinates.

Definition 1 may use another norm for RGB images and mentions only 8 directions d for simplicity of the current implementation, The derivatives in (3.1) can be computed for any direction d. For a fixed directional vector d, consider all straight lines L given by $d_x x + d_y y + t = 0$ with the gradient d such that the shift t takes all integer values when the line intersects the image $\Omega = [0,w] \times [0,h]$.

We select segments $S \subset L$ such that the contrast function f_L over S has persistently larger values than over the rest of L. Here are the steps of Stage 1.

Step (1a). After Gaussian filtering an image I, compute the image gradient DI using 2×2 masks in (2.2). Any more advanced de-noising is possible. One can consider more sophisticated estimates of DI instead of 2×2 masks in (2.2).

Step (1b). For every line L parallel to one of 8 directions d and an integer point $(x,y) \in \Omega$ estimate the derivative of I in the direction orthogonal to d by (3.1).

The naive edge detection in the discrete setting can actually stop at this stage and output all points whose gradient magnitudes are above a certain threshold.

3.2   Stage 2: Finding Strongest Segments by Their Persistence

The aim of this Stage 2 is to automatically select one or several segments within a fixed line L that well approximate contours of an image I within L.

Stage 1 has essentially reduced the detection problem from dimension 2 to 1. Indeed, the input for Stage 2 is a graph of the contrast function $f_L : L \to \mathbb{R}$

sampled at integer points in the line L. The output will be segments $S_1, \ldots, S_k \subset L$ over which the function f is substantially larger than over the rest of L.

The traditional approach is to manually choose a contrast threshold u and consider line segments where the contrast is sufficiently high: $f \geq u$.

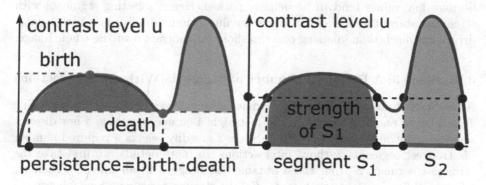

Fig. 4. Segments in superlevel sets $f_L^{-1}[u, +\infty)$ of a contrast function f_L grow and merge when the contrast level u goes down. The *strength* of a segment S is $\int_S f(p)dp$.

The new approach is very different and has no thresholds at this stage. Following the key idea of Topological Data Analysis, we consider the sequence of all superlevel sets $f_L^{-1}[u, +\infty)$ when the level u goes down from a global maximum to a reasonable minimum. During this evolution, connected components of $f_L^{-1}[u, +\infty)$ appear at local maxima of f, grow and merge into larger components. Figure 4 shows two segments that merge into a longer one.

Definition 2. *The* birth *of each component (line segment S) is the maximum value of f_L over S. The* death *of S is the level where S merges with another component. By the standard* elder rule *of persistence [7, p. 150], the older component (with a larger birth here) survives and the younger one dies. The whole process can be combinatorially described by a* topological barcode *of intervals $(death, birth]$ or a* persistence diagram *of pairs $(birth, death)$.*

The main advantage of the persistence diagram is the stability under bounded noise. If a function f_L is perturbed up to ϵ (say with respect to the L_∞ norm), the diagram is perturbed also up to ϵ with respect to the so-called bottleneck distance [5]. Since outliers may destroy this stability we suggest a new measure for selecting segments by analyzing the sequence of superlevel sets.

Definition 3. *At every fixed level u, any current segment (a connected component of $f_L^{-1}[u, +\infty)$) has the* strength *$|S| = \int_S f_L(p)dp$, which is approximated for a pixel-based image as the sum of $f_L(p)$ for all points $p \in S$ with integer coordinates. Figure 4 shows the strength $|S|$ as the area below the graph of f_L.*

Now all segments at the fixed level u can be ranked according to their strengths, say $S_1 > \cdots > S_k$. To separate strongest segments from the rest, below we use the heuristic of the widest gap between these ordered strengths.

Find an index i such that the difference $S_i - S_{i+1}$ (the gap between successive strengths) is largest over all $i = 1, \ldots, k - 1$. The segments with the strengths $S_1, \ldots, S_{i(u)}$ above this widest gap are called strongest at the current level u.

Contrast values of real images have wide gaps usually in a high-value range, because low values tend to be densely packed. Hence selecting segments with strengths above the widest gap (in every line L individually) is a better data-driven approach than guessing one threshold for contrast over the whole image.

3.3   Stage 3: A Required Number of Segments Without Intersections

After Stage 2 above we have one or more strongest segments within every line L parallel to one of 8 directions. So a straight line may continue a few disjoint segments, not necessarily one. Final Stage 3 greedily selects a required number of strongest segments without intersections. In more details, we first take the strongest segment S from those obtained at Stage 2 in all lines L. Then we remove all line segments that *contradict* the strongest segment S as follows.

Definition 4. *A line segment S' contradicts another line segment S if either*
(4a) S' is parallel to S and is away from S within 3 pixels (a default value) or
(4b) S' intersects the segment S, endpoints of S can be inside S' and vice versa.

The default value of 3 pixels between line segments is the reasonable minimum, because the accuracy of human-drawn contours in the BSD is 2 pixels. After removing the chosen segment S_1 all segments contradicting S_1, we select the strongest segment S_2 from the remaining ones, again remove all segments contradicting S_2 and so on until we have found a required number of segments or there are no segments left from Stage 2.

To quickly check the conditions of Definition 4, we keep all segments parallel to one of 8 directions d in a binary tree T_d ordered by the following *identifier* of a line parallel to d. This tree is implemented as a multi-map structure of pairs (identifier of a line L, a segment S within L).

For any non-horizontal infinite line L, this identifier is the x-coordinate at the intersection of L with the x-axis. For a horizontal line L parallel to $d = (1, 0)$, the identifier of L is the constant y-coordinate of L.

Since the number k of required segments is usually much smaller than the number n of pixels, Theorem 5 justifies that the PLSD algorithm is near linear.

Theorem 5. *For any image consisting of n pixels, the algorithm PLSD outputs k straight line segments in time $O(kn \log n)$ and requires $O(n)$ space.*

Proof. For an image of $n = w \times h$ pixels and any of the 8 basic directions d, there at most $w + h = O(\sqrt{n})$ straight lines L parallel to d. Each of these lines contains at most $w + h = O(\sqrt{n})$ points (x, y) with integer coordinates.

For every fixed line L, we use a union-find structure to analyze the evolution of segments $S \subset L$, which are connected components of superlevel sets $f_L^{-1}[u, +\infty)$. We sort the contrast values of $O(\sqrt{n})$ points p within the line L in time $O(\sqrt{n}\log n)$ and process them starting from the largest.

All current segments are kept in a binary tree of size $O(\sqrt{n})$. When a new point p is added to a superlevel set $f_L^{-1}[u, +\infty)$, we have one of three cases:

(5a) p forms a new segment consisting of a single node.
(5b) p joins one of existing straight line segments S.
(5c) p is the merge point of two segments S_1 and S_2.

In case (5a) a new segment is added to the binary tree T_d in time $O(\log n)$. In case (5b) the existing segment is found and its strength is updated in time $O(\log n)$. In case (5c) two segments are removed and a new larger one is inserted in time $O(\log n)$. In general, making $O(\sqrt{n})$ updates, the union-find structure [18] maintains connected components of $O(\sqrt{n})$ points $p \in L$ in time $O(\sqrt{n}\log n)$.

At every step of Stage 2 we update the binary tree of $O(\sqrt{n})$ segments (parallel to a fixed direction d) as we need to know the widest gap between successive strengths. The binary trees T_d ordered by unique identifiers of lines L parallel to d help to remove in time $O(\sqrt{n}\log n)$ all segments contradicting a current strongest segment in the sense of Definition 4. The factor k in the complexity is from the number of strongest segments that are searched in the trees T_d. $\square$

4   Experiments on 500 BSD Images

4.1   The Boundary Recall Benchmark BR(2) from BSD500

The Berkeley Segmentation Database (BSD) [1] consists of 500 images widely used for evaluating segmentation algorithms due to human-sketched ground truth boundaries. The human-drawn boundaries for each image are discretized and saved as a set G of ground-truth pixels. If E is another set of pixels produced by an edge detection algorithm, the standard *Boundary Recall* is

$$BR(G, E, \epsilon) = \frac{\#\{\text{pixels } p \in G \ : \ \text{distance}(p, E) \leq \epsilon\}}{|G|},$$

where $\text{distance}(p, E)$ is the Euclidean distance between (centers of) p and its closest neighbor in E. The standard offset of $\epsilon = 2$ pixels for the Boundary Recall is usually chosen, because human drawings cannot be more accurate.

Since there are up to 7 human-drawn boundaries B per image, the convention is to compute $BR(E, \epsilon)$ for a fixed image as the maximum of $BR(G, E, \epsilon)$ over all ground-truths G, hence over the best human drawing. The final Boundary Recall $BR(\epsilon)$ in Fig. 8 is the average of $BR(E, \epsilon)$ over all 500 images.

For any line detector at subpixel resolution, there is little sense to discretize its output set S of line segments. We compute the Euclidean distance (p, S) from a ground truth pixel $p \in G$ to a closest line segment in the output S.

$$BR(G, S, \epsilon) = \frac{\#\{\text{pixels } p \in G \ : \ \text{distance}(p, S) \leq \epsilon\}}{|G|}.$$

4.2   LSDA vs PLSD on the Boundary Recall BR(2)

Since the LSDA was extensively compared with past line segment detections in [10, Sect. 4], this paper quantitatively compares the PLSD only with the LSDA. The LSDA "was designed as an automatic image analysis tool and must work without requiring any parameter tuning" [10, p. 2]. We followed the advice of Grompone von Gioi [10] to run the LSDA with the default parameters. Hence the LSDA results are represented by a single black dot whose horizontal coordinate is the average number of line segments across BSD500.

For each BSD image, the LSDA produced a number of line segments according to the "a contrario" model. The PLSD algorithm is more flexible and can output a smaller or larger number of segments by users' choice, see Figs. 5, 6 and 7.

For a fair comparison, on every image we first ran the LSDA code and then asked PLSD to output the same number of line segments as LSDA. However, in some cases the PLSD algorithm outputs a smaller number of edges because all edges are required to be non-intersecting.

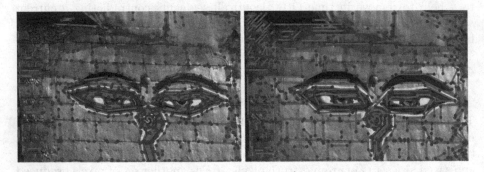

Fig. 5. Left: LSDA output on image 56028 in BSD. **Right**: More and longer straight line segments are found by PLSD.

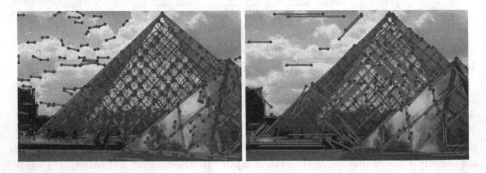

Fig. 6. Left: the LSDA output has too short segments in image 223060 from the BSD500. **Right**: many longer line segments are found by the new PLSD algorithm.

Fig. 7. Left: the LSDA missed some vertical lines in image 5096 from the BSD500. **Right**: many longer line segments are found by the new PLSD algorithm.

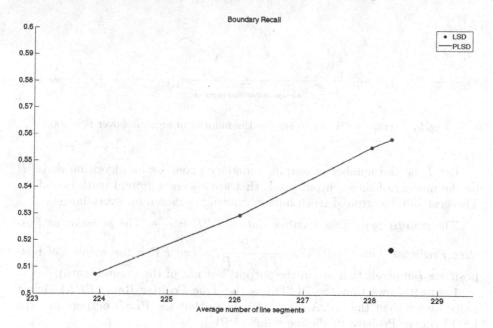

Fig. 8. Boundary Recall BR(2) for PLSD and LSDA on 500 BSD images, the horizontal axis shows the average number of segments.

The graph in Fig. 8 shows one black dot for LSDA and the blue polygonal curve with 4 dots corresponding to offset = 2, 3, 4, 5 pixels (from right to left). This offset parameter is used in condition (4) to avoid very close parallel segments.

Any straight line segment found by LSDA or PLSD is discretized by drawing black lines on a white background and extracting resulting black pixels as the discrete output, because the human drawings were discretized in BSD.

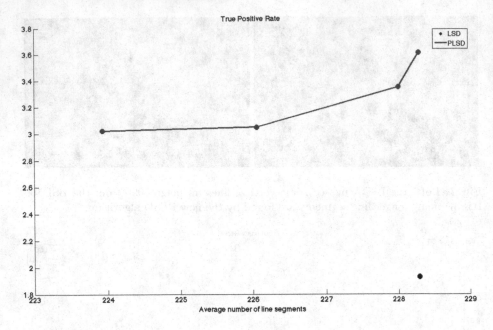

Fig. 9. Average TPR in percents vs the number of segments over BSD500.

Let P be the number of output boundary pixels by an algorithm and TP be the number of those output pixels that are also in a ground truth boundary. The best suitable ground truth human drawing is chosen for every image.

The *sensitivity* or *True Positive Rate* is $TPR = \dfrac{TP}{P}$. The *precision* or *Positive Predictive Value* is $PPV = \dfrac{TP}{TP+FP}$, where FP is the number of false positives (all pixels that are in the output, but not in the ground truth).

Figure 9 shows that the PLSD has the True Positive Rate (TPR) almost twice better than the LSDA. Figure 10 shows that the PLSD outperforms the LSDA on the Positive Predictive Value (PPV).

5   Discussion and Conclusions

The experiments in Sect. 4 have demonstrated that the proposed detection of line segments parallel to one of 8 directions already outperforms the state-of-the-art algorithm that allows line segments with any slope. The data-driven approach of selecting strongest segments can be extended to more than 8 directions.

Other possible improvements are better filtering, e.g. optimizing the size and sigma in the Gaussian kernel, and more advanced de-noising before Step (3.1a). The current non-optimized code runs for about 1 sec per BSD image on a laptop with 8Gb Ram, which is a bit slower than the LSDA on the same machine.

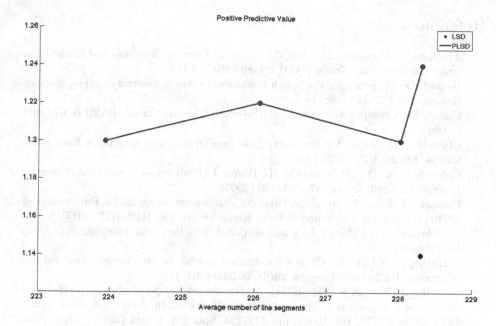

Fig. 10. Average PPV in percents vs the number of segments over BSD500.

The straight line segments can be used as very economical descriptors of complicated scenes. For example, training convolution neural networks on straight line sketches can be much faster than on original images.

The novel method of automatic selection in Subsect. 3.2 can be used for finding skeletons of objects [4,13,15,16], where thresholds should be avoided. Here is the summary of contributions to the line segment detection problem.

- The PLSD allows a user to fix a desired number of strongest line segments.
- All line segments in the output have no intersections by Definition 4, hence PLSD can be easily extended to planar skeletons and polygonal meshes.
- The PLSD has a near linear computational complexity by Theorem 5.
- A thresholding of contrast values was avoided due to the new data-driven method motivated by a multi-scale approach of Topological Data Analysis.
- The PLSD has outperformed the LSDA on the Boundary Recall, e.g. for the default offset of 3 pixels the measure BR(2) has improved from 0.517 to 0.559.

Acknowledgments. We thank all reviewers for helpful suggestions. This work was supported by the EPSRC grant "Application-driven Topological Data Analysis" (EP/R018472/1).

References

1. Arbelaez, P., Maire, M., Fowlkes, C., Malik, J.: Contour detection and hierarchical image segmentation. Trans. PAMI **33**, 898–916 (2011)
2. Ballard, D.: Generalizing the hough transform to detect arbitrary shapes. Pattern Recogn. **13**, 111–122 (1981)
3. Canny, J.: A computational approach to edge detection. Trans. PAMI **8**, 679–698 (1986)
4. Chernov, A., Kurlin, V.: Reconstructing persistent graph structures from noisy images. Image-A **3**, 19–22 (2013)
5. Cohen-Steiner, D., Edelsbrunner, H., Harer, J.: Stability of persistence diagrams. Discrete Comput. Geom. **37**, 103–130 (2007)
6. Duan, L., Lafarge, F.: Image partitioning into convex polygons. In: Proceedings of CVPR (Computer Vision and Pattern Recognition), pp. 3119–3127 (2015)
7. Edelsbrunner, H., Harer, J.: Computational Topology: An Introduction. AMS, Providence (2010)
8. Forsythe, J., Kurlin, V.: Convex constrained meshes for superpixel segmentations of images. J. Electron. Imaging **26**(6), 061609 (2017)
9. Forsythe, J., Kurlin, V., Fitzgibbon, A.: Resolution-independent superpixels based on convex constrained meshes without small angles. In: Bebis, G., et al. (eds.) ISVC 2016. LNCS, vol. 10072, pp. 223–233. Springer, Cham (2016). https://doi.org/10.1007/978-3-319-50835-1_21
10. Grompone von Gioi, R.: A Contrario Line Segment Detection. Briefs in Computer Science. Springer, New York (2014). https://doi.org/10.1007/978-1-4939-0575-1
11. Grompone von Gioi, R., Jakubowicz, J., Morel, J.M., Randall, G.: LSD: a line segment detector. Image Process. Line **2**, 35–55 (2012)
12. Kahn, P., Kitchen, L., Riseman, E.: A fast line finder for vision-guided robot navigation. Trans. PAMI **12**, 1098–1102 (1990)
13. Kalisnik, S., Kurlin, V., Lesnik, D.: A high-dimensional homologically persistent skeleton. Adv. Appl. Math. **102**, 113–142 (2019)
14. Kurlin, V.: Auto-completion of contours in sketches, maps and sparse 2D images based on topological persistence. In: Proceedings of SYNASC 2014 Workshop CTIC: Computational Topology in Image Context, pp. 594–601. IEEE (2014)
15. Kurlin, V.: A homologically persistent skeleton is a fast and robust descriptor of interest points in 2d images. In: Azzopardi, G., Petkov, N. (eds.) CAIP 2015. LNCS, vol. 9256, pp. 606–617. Springer, Cham (2015). https://doi.org/10.1007/978-3-319-23192-1_51
16. Kurlin, V.: A one-dimensional homologically persistent skeleton of a point cloud in any metric space. Comput. Graph. Forum **34**, 253–262 (2015)
17. Kurlin, V.: A fast persistence-based segmentation of noisy 2d clouds with provable guarantees. Pattern Recogn. Lett. **83**, 3–12 (2016)
18. Tarjan, R.: Data Structures and Network Algorithms. SIAM, Philadelphia (1983)

On the Nonlinear Statistics of Optical Flow

Henry Adams$^{(\boxtimes)}$, Johnathan Bush, Brittany Carr, Lara Kassab,
and Joshua Mirth

Colorado State University, Fort Collins, CO 80523, USA
{adams,bush,carr,kassab,mirth}@math.colostate.edu

Abstract. In *A naturalistic open source movie for optical flow evaluation*, Butler et al. create a database of ground-truth optical flow from the computer-generated video *Sintel*. We study the high-contrast 3×3 patches from this video, and provide evidence that this dataset is well-modeled by a *torus* (a nonlinear 2-dimensional manifold). Our main tools are persistent homology and zigzag persistence, which are popular techniques from the field of computational topology. We show that the optical flow torus model is naturally equipped with the structure of a fiber bundle, which is furthermore related to the statistics of range images.

Keywords: Optical flow · Computational topology
Persistent homology · Fiber bundle · Zigzag persistence

1   Introduction

A video records a moving three-dimensional world as a sequence of two-dimensional images. The apparent motion of the two-dimensional images, due to changing brightness, is called *optical flow*. The optical flow at a frame is a vector field, where the vector at each pixel points to where that pixel appears to move for the subsequent frame [10].

A fundamental problem is to estimate optical flow from a video sequence [7,23]. It is impossible to recover the optical flow field exactly using only a video sequence; for example, if one is given a video of a spinning barber's pole, one does not know (without prior knowledge) whether the pole is moving up or instead spinning horizontally. Given these difficulties, algorithms estimating optical flow must exploit or make assumptions about the statistics of optical flow, and hence there is interest in understanding these statistics.

As no instrument measures ground-truth optical flow, databases must be generated. One example database is from the 3-D, animated, open source, short film Sintel, which has several desirable features. The scenes are long, and the movements and textures are more complex than in some of the other datasets. Since the film is open source, the optical flow data is available for analysis (see Fig. 1). The Sintel optical flow dataset is described in detail in [14].

© Springer Nature Switzerland AG 2019
R. Marfil et al. (Eds.): CTIC 2019, LNCS 11382, pp. 151–165, 2019.
https://doi.org/10.1007/978-3-030-10828-1_12

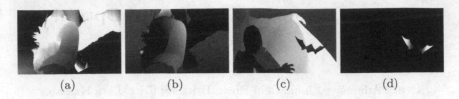

(a)   (b)   (c)   (d)

Fig. 1. Two sample optical flows extracted from the Sintel database. Horizontal components in (a) and (c); vertical components in (b) and (d). White corresponds to flow in the positive direction (right or up) and black corresponds to the negative direction.

In this paper we use the topological machinery of [2,17] to study the nonlinear statistics of optical flow from the Sintel dataset. First, we build a space of high-contrast 3×3 optical flow patches. Using Vietoris–Rips complexes and persistent homology, we identify the topologies of dense subsets of this space. The densest patches lie near a circle, the *horizontal flow circle* [1]. Then, in a more refined analysis, we select out the optical flow patches whose predominant direction of flow is a small bin of angle values. We show that the patches in each such bin are well-modeled by a circle; each such circle is explained by the nonlinear statistics of range image patches. We show that these circles at each angle piece together, via the structure of a *fiber bundle*, into a torus model for optical flow.

The torus model for the nonlinear statistics of optical flow could also be used for optical flow compression. Indeed, one can express a 3×3 optical flow patch as an average flow vector, plus a patch on a 2-dimensional torus, plus a 3×3 error vector whose entries will tend to be small in magnitude.

We survey related work in Sect. 2, and in Sect. 3 we introduce our topological methods. We describe the spaces of high-contrast optical flow patches in Sect. 4, and present our main results in Sect. 5. Our code is available at https://bitbucket.org/Cross_Product/optical_flow/.

2   Related Work

2.1   Optical Flow Datasets

There are a variety of databases that reconstruct ground truth optical flow samples. The Middlebury dataset in [5] ranges from real stereo imagery of rigid scenes to realistic synthetic imagery; the database contains public ground truth optical flow training data along with sequestered ground truth data for the purpose of testing algorithms. The data from [30] consists of twenty different synthetic scenes with the camera and movement information provided. The KITTI Benchmark Suite [24] uses a car mounted with two cameras to film short clips of pedestrians and cars; attached scanning equipment allows one to reconstruct the underlying-truth optical flow for data testing and error evaluation.

Another example is the database created by Roth and Black [33] to study the statistics of optical flow. Unlike databases used to test optical flow estimation, the Roth and Black database does not include accompanying video sequences. Freed from this constraint, Roth and Black generate optical flow for a wide variety of natural scenes by pairing range images[1] with camera motions. The resulting optical flow can be calculated from the geometry of the static scene and of the camera motion. The database includes only optical flow from static scenes seen by a moving camera: no objects in the field of view move independently.

By contrast, the ground-truth Sintel optical flow database [14], which we study in this paper, is computed directly from the film's motion vectors; there is no need to reconstruct the flow via computations from laser scans.

2.2 Optical Flow Applications

Optical flow estimation is most commonly used in computer vision tasks. Computer vision is a process wherein a computer takes in visual data, analyzes the data via various statistics (for example after estimating optical flow), and then outputs information or a decision based on the data. Computer vision algorithms utilizing the estimation of optical flow can be found in facial recognition software [6], in driving autonomous cars [24], and in robotic tracking [26].

2.3 Optical and Range Image Work from the Topological Perspective

We briefly describe a subset of papers that have analyzed the statistics of optical images, range images, and optical flow from the perspective of computational topology. Foundational papers in this area include [28], which proposes a circular model for 3×3 optical image patches, and [17], which uses persistent homology to extend this circular model to both a three-circle model and a Klein bottle model for different dense core subsets.

The nonlinear statistics of range image patches (which contain a distance at each pixel) will play an important role in our work. In [28], the authors observe that high-contrast 3×3 range patches from [27] cluster near binary patches. The paper [2] uses persistent homology to find that the densest range clusters are arranged in the shape of a circle. After enlarging to 5×5 or 7×7 patches, the entire primary circle in Fig. 2(a) is dense. The patches forming the range primary circle are binary approximations to linear step edges; see Fig. 2(b).

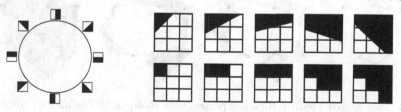

Fig. 2. (Left) Range patch primary circle [2]. White regions are far; black regions are near. (Right) The top row contains linear step edges; the bottom row contains their range image binary approximations.

---

[1] A range image contains a distance at each pixel.

The paper [1] uses the nudged elastic band method to propose the horizontal flow circle model for (horizontal) optical flow patches.

3   Topological Machinery

In this section we describe how to use only a finite sampling from some unknown underlying space to estimate the underlying space's topology. The first step is to build a nested family of simplicial complexes, and the second is to apply persistent homology. This is the same topological approach used to analyze optical and range image patches in [2,17]. We refer the interested reader to [4,25] for more information on homology, to [15,21,22,37] for introductions to persistent homology, and to [3,8,11,12,19,29,34–36] for example applications of persistent homology to sensor networks, machine learning, biology, medical imaging, etc.

3.1   Vietoris–Rips Complexes

Our nested complexes will be Vietoris–Rips simplicial complexes. The main idea is to define all data points to be vertices of the complex, and to define a simplex σ on each finite set of vertices within a given diameter. Indeed, let (X, d) denote a metric space, and fix a scale parameter $r \geq 0$. The *Vietoris–Rips simplical complex* with vertex set X and scale parameter r, denoted $\mathrm{VR}(X; r)$, is defined as follows. A finite subset $\sigma = \{x_1, \ldots, x_n\} \subseteq X$ is a face of $\mathrm{VR}(X; r)$ whenever $\mathrm{diam}(\sigma) \leq r$ (i.e., whenever $\sup_{1 \leq i \leq j \leq n}\{d(x_i, x_j)\} \leq r$). By definition, $\mathrm{VR}(X; r) \subseteq \mathrm{VR}(X; r')$ whenever $r \leq r'$, so this family is indeed nested.

Let us consider an example. Let X be 21 points which (unknown to us) are sampled with noise from a circle. Figure 3 contains four nested Vietoris–Rips complexes built from X, with r increasing from left to right. The black dots denote X. In (a), r is small enough that a loop has not yet formed. In (b), r is such that we recover instead a figure-eight. In (c), $\mathrm{VR}(X; r)$ recovers a circle. In (d), r is large enough that the loop has filled to a disk.

Fig. 3. Four nested Vietoris–Rips complexes, with β_0 equal to 1 in all four complexes, and with β_1 equal to 0, 2, 1, and 0.

3.2   Persistent Homology

Betti numbers are one way of distinguishing between different topological spaces: a necessary condition for two spaces to be homotopy equivalent is for all of their Betti numbers to be equal. The k-th Betti number of a topological space, denoted β_k, is the rank of the k-th homology group. Roughly speaking, β_k is the number of "k-dimensional holes" in a space, where the number of 0-dimensional holes is the number of connected components. For an n-dimensional sphere with $n \geq 1$, we have $\beta_0 = 1$ and $\beta_n = 1$.

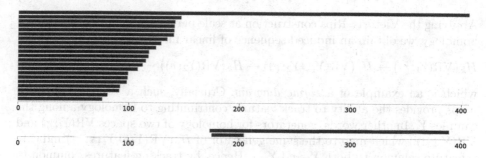

Fig. 4. (Top) The 0-dimensional persistence barcode associated to the dataset in Fig. 3. (Bottom) The 1-dimensional persistence barcode associated to the same dataset.

If we want to estimate the topology of the underlying space by the topology of $\text{VR}(X;r)$, the choice of r is important. However, without knowing the underlying space, we do not know how to make this choice. Hence, we use persistent homology [22,37], which allows us to compute the Betti numbers over a range of r-values and to display the result as a persistent homology barcode. See Fig. 4.

Persistent homology depends on the the fact that the map from a topological space Y to its k-th homology group $H_k(Y)$ is a functor. This means that for $r \leq r'$, the inclusion $\text{VR}(X;r) \hookrightarrow \text{VR}(X;r'))$ of topological spaces induces a map $H_k(\text{VR}(X;r)) \to H_k(\text{VR}(X;r'))$ between homology groups [21].

The horizontal axis in Fig. 4 contains the varying r-values. At a given scale r, the Betti number β_k is the number of intervals in the dimension k plot that intersect the vertical line through scale r. In the dimension 0 plot, we see the 21 disjoint spaces joining into one connected component as r increases. The two intervals in the dimension 1 plot correspond to the two loops that appear: each interval begins when a loop forms and ends when that loop fills to a disk.

The topological profile of this example, $\beta_0 = 1$ and $\beta_1 = 1$, is obtained for a long range of r-values in Fig. 4. The idea of persistent homology is that long intervals in the persistence barcodes correspond to real topological features of the underlying space. We often disregard short intervals as noise. Hence, this barcode reflects the fact that our points X were noisily sampled from a circle.

3.3 Zigzag Persistent Homology

Zigzag persistence [16,18] provides a generalization of the theory of persistent homology. In zigzag persistence, the direction of maps along a sequence of topological spaces is arbitrary, as opposed to the unidirectional sequence of maps in persistent homology. Given a large dataset Y, one may attempt to estimate the topology of Y by instead estimating the topology of a number of smaller subsets $Y_i \subseteq Y$. Toward that end, consider the following diagram of inclusion maps between subsets of the data.

$$Y_1 \hookrightarrow Y_1 \cup Y_2 \hookleftarrow Y_2 \hookrightarrow Y_2 \cup Y_3 \hookleftarrow Y_3 \hookrightarrow \cdots \hookleftarrow Y_n. \tag{1}$$

Applying the Vietoris–Rips construction at scale parameter r and k-dimensional homology, we obtain an induced sequence of linear maps

$$H_k\big(\mathrm{VR}(Y_1;r)\big) \to H_k\big(\mathrm{VR}(Y_1 \cup Y_2;r)\big) \leftarrow H_k\big(\mathrm{VR}(Y_2;r)\big) \to \cdots \leftarrow H_k\big(\mathrm{VR}(Y_n;r)\big)$$

which is an example of a *zigzag diagram*. Crucially, such a sequence of linear maps provides the ability to track features contributing to homology among the samples Y_i. In other words, generators for homology of two spaces $\mathrm{VR}(Y_i;r)$ and $\mathrm{VR}(Y_{i+1};r)$ which map to the same generator of $H_k\big(\mathrm{VR}(Y_i \cup Y_{i+1};r)\big)$ indicate a feature common to both Y_i and Y_{i+1}. Hence, by tracking features common to all samples Y_i, one may estimate the topology of Y without explicitly computing the persistent homology of the entire dataset.

3.4 Fiber Bundles

Our identification of a torus model for the MPI-Sintel optical flow dataset is guided by the notion of a fiber bundle. Precisely, a fiber bundle is a tuple (E, B, f, F), where E, B, and F are topological spaces and $f \colon E \to B$ is a continuous map satisfying a so-called *local triviality* condition, described below. We call B the *base space*, E the *total space*, and F the *fiber*. The local triviality condition on f is as follows: given $b \in B$, there exists an open set $U \subseteq B$ containing b and a homeomorphism $\varphi \colon f^{-1}(U) \to U \times F$ such that $\mathrm{proj}_U \circ \varphi = f|_{f^{-1}(U)}$, where proj_U denotes the projection onto the U–component. In other words, we require $f^{-1}(U)$ to be homeomorphic to $U \times F$ in a particular way. Therefore, for any $p \in B$, we have $f^{-1}(\{p\}) \cong F$. *Locally*, the total space E looks like $B \times F$, while *globally*, a fiber bundle contains information about how these copies of the fiber F may be "twisted".

As an example, both the cylinder and the Möbius band may be realized as fiber bundles with base space the circle S^1, and with fibers the unit interval $[0, 1]$. In the case of the Möbius band, the global structure of the fiber bundle gives a "half twist" as one loops around the circle, whereas the global structure of the cylinder does not contain a twist. Locally, however, both spaces look the same, as each have the same fiber above each point of S^1.

Analogously, both the torus and the Klein bottle may be realized as fiber bundles over S^1, with fibers homeomorphic to S^1. In this case, the fibers of the Klein bottle "twist" in a particular way, whereas the fibers of the torus do not.

In Sect. 5, we use persistent homology to provide evidence that the MPI-Sintel dataset is naturally equipped with the structure of a fiber bundle over a circle, with each fiber being a circle. It is not a priori clear whether this fiber bundle model should be the orientable torus or the nonorientable Klein bottle; indeed, the space of optical image patches as studied in [17] is well-modeled by a Klein bottle. However, in Sect. 5.2, we provide evidence that this optical flow fiber bundle is a torus.

4   Spaces of Flow Patches

The *MPI-Sintel* optical flow dataset [14] contains 1041 optical flow fields, each 1024×436 pixels. The dataset originates from the open-source animated film *Sintel* [32], which contains a variety of realism-enhancing effects, including widely varied motion, illumination, and blur. This data is extracted from 23 scenes of indoor and outdoor environments, with up to 49 frames per scene.

We create spaces of high-contrast optical flow patches, $X(k, p)$ or $X_\theta(k, p)$. The version $X_\theta(k, p)$ includes only those optical flow patches whose predominant angle is near $\theta \in [0, \pi)$. Our preprocessing is similar to that done in [2,17,28].

Step 1: We randomly choose $4 \cdot 10^5$ size 3×3 optical flow patches from the MPI-Sintel database. Each patch is a matrix of ordered pairs, where u_i and v_i are the horizontal and vertical components of the flow vector at pixel i.

$$\begin{bmatrix} (u_1, v_1) & (u_4, v_4) & (u_7, v_7) \\ (u_2, v_2) & (u_5, v_5) & (u_8, v_8) \\ (u_3, v_3) & (u_6, v_6) & (u_9, v_9) \end{bmatrix}$$

For convenience, we rearrange each patch x to be a length-18 vector, $x = (u_1, \ldots, u_9, v_1, \ldots, v_9)^T \in \mathbb{R}^{18}$. We define u and v to be the vectors of horizontal and vertical flow: $u = (u_1, u_2, \ldots, u_9)^T$ and $v = (v_1, v_2, \ldots, v_9)^T$.

Step 2: Let $i \sim j$ denote that pixels i and j are adjacent in the 3×3 patch. For each patch x, we compute the contrast norm $\|x\|_D$ by summing the squared differences between all adjacent pixels and then taking the square root, namely:

$$\|x\|_D^2 = \sum_{i \sim j} \|(u_i, v_i) - (u_j, v_j)\|^2 = \sum_{i \sim j} (u_i - u_j)^2 + (v_i - v_j)^2 = u^T D u + v^T D v.$$

Matrix D, which stores the adjacency information of the pixels in a 3×3 patch, is a symmetric positive definite 9×9 matrix given in [28].

Step 3: We select out the patches that have a contrast norm in the top 20% of the entire sample; hence we only studying high-contrast flow patches, which we expect to follow a different distribution than low-contrast patches. After doing so, we replace each patch x with its contrast-normalized patch $x/\|x\|_D$; this places each patch on the surface of an ellipsoid. We need not worry about dividing by contrast norm zero, as such patches are not high-contrast.

Step 4: We further normalize the patches to have zero average flow. For a patch x, let $\bar{u} = \frac{1}{9}\sum_{i=1}^{9} u_i$ be the average horizontal flow, and let $\bar{v} = \frac{1}{9}\sum_{i=1}^{9} v_i$ be the average vertical flow. We replace each contrast-normalized vector x with $(u_1 - \bar{u}, \ldots, u_9 - \bar{u}, v_1 - \bar{v}, \ldots, v_9 - \bar{v})^T$. The purpose of studying mean-centered optical flow patches is that one can represent any optical flow patch as its mean vector plus a mean-centered patch.

Step 5: If we are computing $X_\theta(k, p)$ (as opposed to $X(k, p)$), then we compute the predominant direction of each mean-centered flow patch, as follows. For each 3×3 patch, construct a 9×2 matrix X whose i-th row is $(u_i, v_i) \in \mathbb{R}^2$. We apply principal component analysis (PCA) to X in order to retrieve the principal component with the greatest component variance (i.e., the direction that best approximates the deviation from the mean). The angle of this line (in $[0, \pi)$ or $\mathbb{RP}^1$) is defined to be the predominant direction of the patch. We select out only those patches whose predominant direction is in the range of angles on $\mathbb{RP}^1 = [0, \pi)$ from $\theta - \frac{\pi}{12}$ to $\theta + \frac{\pi}{12}$.

Step 6: If we have more than 50,000 patches, then we randomly subsample down to 50,000 random patches for the sake of computational feasibility.

Step 7: We now have at most 50,000 high-contrast normalized optical flow patches. Instead of trying to approximate the topology of such a diverse space, we restrict to dense core subsets. We use the density estimator ρ_k, where $\rho_k(x)$ is the distance from x to its k-th nearest neighbor. Note that ρ_k is inversely proportional to density. Decreasing (or increasing) the choice of k produces a more local (or global) estimate of density. We select out the top $p\%$ densest points, based on the density estimator ρ_k. We denote this set of patches by $X(k, p)$ (or by $X_\theta(k, p)$, in the case where Step 5 is performed).

5   Results

5.1   The Horizontal Flow Circle

The dataset $X(300, 30)$ is well-modeled by the horizontal flow circle, as identified in [1] via the nudged elastic band method. We locate this circle by projecting onto suitable basis vectors. Let $e_1, e_2, \ldots, e_8$ be the discrete cosine transform (DCT) basis for 3×3 scalar patches, normalized to have mean zero and contrast norm one. This basis is given in [28]. For convenience, we rearrange each e_i to be a length-9 vector. Now, let z be the length-9 zero vector. For each $i = 1, 2, \ldots, 8$, we define optical flow vectors $e_i^u = \begin{pmatrix} e_i \\ z \end{pmatrix}$ and $e_i^v = \begin{pmatrix} z \\ e_i \end{pmatrix}$. Note that $e_i^u, e_i^v \in \mathbb{R}^{18}$ correspond respectively to optical flow in the horizontal and vertical directions. We change coordinates from the canonical basis for $\mathbb{R}^{18}$ to the 16 basis vectors $e_1^u, \ldots, e_8^u, e_1^v, \ldots, e_8^v$. Some of these basis vectors are in Fig. 6 (left). The projection of $X(300, 30)$ onto basis vectors e_1^u and e_2^u, shown in Fig. 5 (left), reveals the circular topology.

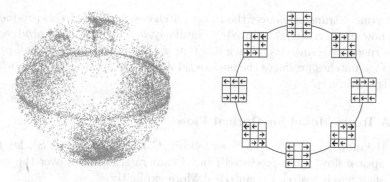

Fig. 5. (Left) Projection of $X(300, 30)$ onto e_1^u and e_2^u. (Right) The horizontal flow circle. The patch at angle α is $\cos(\alpha)e_1^u + \sin(\alpha)e_2^u$.

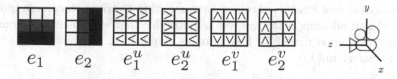

Fig. 6. (Left) In the e_1 and e_2 DCT patches, white pixels are positive and black negative. The arrows in the flow patches e_1^u, e_2^u, e_1^v, and e_2^v show the optical flow vector field patch. (Right) Camera axes.

Let S^1 denote the interval $[0, 2\pi]$ with endpoints identified. The patches in $X(300, 50)$ lie near $\{\cos(\alpha)e_1^u + \sin(\alpha)e_2^u \mid \alpha \in S^1\}$, which we call the *horizontal flow circle*. We will use the statistics of both the camera motion database and the range image database to explain why the horizontal circle is high-density.

Camera motion can be decomposed into six sub-motions. The first three are translation in the x, y, or z direction, commonly referred to as right-left, up-down, or inward-outward translation. The remaining three are rotation about the x, y, or z axis, commonly referred to as pitch, yaw, or roll. See Fig. 6 (right). We will refer to $\theta \in S^1$ camera translation, by which we mean translation of $\cos(\theta)$ units to the right, $\sin(\theta)$ units up, and no units inwards or outwards. Some of the most common camera translations are when $\theta = 0$ or π, i.e. when the camera is translated to the left or right, for example if the camera is mounted on a horizontally moving car or held by a horizontally walking human [33].

In [2] the authors find that high-contrast range patches are dense near the range patch primary circle, defined as $\{\cos(\alpha)e_1 + \sin(\alpha)e_2 \mid \alpha \in S^1\}$ and depicted in Fig. 2(a). Negative and positive pixel coordinates correspond, repectively, to near and far.

Let us consider pairing the common $\theta = 0$ or π camera translations with primary circle range patches. Under camera translation in the xy plane, the flow vector at a foreground pixel has the same direction but greater magnitude than at a background pixel. After the mean-centering normalization in Step 4 of Sect. 4, a $\theta = 0$ camera translation over the range patch $\cos(\alpha)e_1 + \sin(\alpha)e_2$ produces the optical flow patch $\cos(\alpha)e_1^u + \sin(\alpha)e_2^u$. Similarly, $\theta = \pi$ translation produces flow patch $-\cos(\alpha)e_1^u - \sin(\alpha)e_2^u$. Hence $\theta = 0$ or π translation applied to all primary circle range patches produces the horizontal flow circle $\{\cos(\alpha)e_1^u + \sin(\alpha)e_2^u \mid \alpha \in S^1\}$ (Fig. 5 (right)).

5.2   A Torus Model for Optical Flow

Define the map $f : S^1 \times S^1 \mapsto \mathbb{R}^{18}$ as follows. Given $(\alpha, \theta) \in S^1 \times S^1$, let $f(\alpha, \theta)$ be the optical flow patch produced from θ camera translation over the primary circle range patch $\cos(\alpha)e_1 + \sin(\alpha)e_2$. More explicitly,

$$f(\alpha, \theta) = \cos(\theta)\Big(\cos(\alpha)e_1^u + \sin(\alpha)e_2^u\Big) + \sin(\theta)\Big(\cos(\alpha)e_1^v + \sin(\alpha)e_2^v\Big). \qquad (2)$$

The horizontal flow circle is obtained by restricting to common camera motions $\theta \in \{0, \pi\}$ and allowing α to vary. We hypothesize that when neither parameter is restricted, a good model for flow patches is obtained. Hence we ask, what is the image space $\mathrm{im}(f)$?

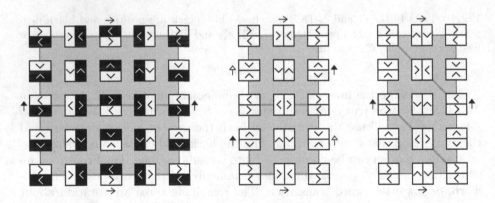

Fig. 7. (Left) The domain of f, namely $\{(\alpha, \theta) \in S^1 \times S^1\}$. (Middle, Right) The flow torus $\mathrm{im}(f)$. The horizontal axis is the angle α, and the vertical axis is the angle θ (respectively $\theta - \alpha$ on the right). The horizontal flow circle is in red. (Color figure online)

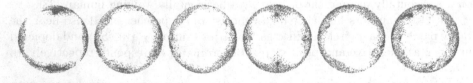

Fig. 8. Consider the set of image patches in $X_\theta(300, 50)$, projected to $\mathbb{R}^2$ via the orthogonal projection onto the basis vectors $\cos(\theta)e_1^u + \sin(\theta)e_1^v$ (horizontal axis) and $\cos(\theta)e_2^u + \sin(\theta)e_2^v$ (vertical axis). The above image shows the resulting plots for $\theta = 0, \frac{\pi}{6}, \frac{2\pi}{6}, \frac{3\pi}{6}, \frac{4\pi}{6}, \frac{5\pi}{6}$. Note that each projection is a circle; we will later show these circles group together to form a torus.

Consider Fig. 7 (left), which shows the domain of f, namely $\{(\alpha, \theta) \in$ 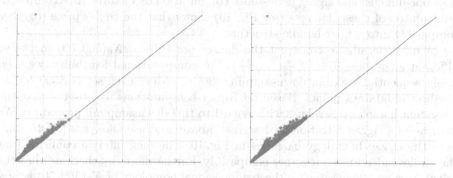 $S^1 \times S^1\}$. This space is a torus, obtained by identifying the outside edges of the figure as indicated by the arrows. A sample patch on this torus is shown in the insert to the right. The black and white rectangles are the foreground and background regions, respectively, of the underlying range patch. Parameter α is the angle of the line separating these regions. The black arrow ($>$, $\vee$, $<$, or $\wedge$) in the white rectangle is the direction θ of camera translation. Together, the black and white arrows show the induced optical flow $f(\alpha, \theta)$. In Fig. 7 (left), parameter α varies in the horizontal direction, and parameter θ varies in the vertical direction.

Fig. 9. The 1-dimensional persistent homology of Vietoris–Rips complexes of $X_\theta(300, 30)$, computed in Ripser [9], confirms that these data sets are well-modeled by a circle (one significant 1-dimensional feature in the top left of each plot). These diagrams contain the same content as persistence intervals, just in a different format: each point is a topological feature with birth scale and death scale given by its x and y coordinates. Above we plot only two sample angles: $\theta = \frac{3\pi}{12}$ (left) and $\theta = \frac{7\pi}{12}$ (right).

Note that for two points (α, θ) and (α', θ') on the torus $S^1 \times S^1$, we have

$$f(\alpha, \theta) = f(\alpha', \theta') \quad \Longleftrightarrow \quad (\alpha, \theta) = (\alpha', \theta') \text{ or } (\alpha, \theta) = (-\alpha', -\theta'). \quad (3)$$

In other words, antipodal points in Fig. 7 (left) produce the same flow patch under the map f. For instance, the horizontal flow circle in red appears twice (note the top and bottom edges are identified).

The image space $\text{im}(f)$ is homeomorphic to the quotient space $\{(\alpha, \theta) \in S^1 \times S^1\}/ \sim$, where $\sim$ denotes the identification $(\alpha, \theta) \sim (-\alpha, -\theta)$. A torus with antipodal points identified remains a torus, and we refer to $\text{im}(f)$ as the *flow torus*. See Fig. 7. The right and left edges of the middle image are identified by shifting one upwards by half its length (not by twisting) before gluing. This suggests a change of coordinates: in Fig. 7 (right) we plot the same flow torus, except we replace the vertical parameter with $\theta - \alpha$. The horizontal flow circle in red now wraps once around one circular direction and twice around the other.

We hypothesize that $\text{im}(f)$, the flow torus, is a good model for high-contrast optical flow. This is confirmed in part by Figs. 8 and 9, which show that for any angle θ, the patches $X_\theta(300, 30)$ (with predominant flow in direction θ) form a circle. Together these circles group together to form a torus, equipped with the structure of a fiber bundle. Indeed, the map from the torus to the predominant angle θ of each patch is a fiber bundle with total space a torus, with base space the circle of all possible predominant angles θ, and with each fiber a circle (arising from the primary circle of range images in Fig. 2).

We do a zigzag persistence computation in order to confirm that the circular fibers glue together to form a torus. Consider the following zigzag diagram.

$$X_0(300, 50) \hookrightarrow X_0(300, 50) \cup X_{\frac{\pi}{12}}(300, 50) \hookleftarrow X_{\frac{\pi}{12}}(300, 50) \hookrightarrow \ldots \hookleftarrow X_{\frac{11\pi}{12}}(300, 50).$$

The one-dimensional zigzag persistence computation of Vietoris–Rips complexes built on top of these datasets (see Fig. 10) shows that the circles piece together compatibly into a fiber bundle structure.

In more detail, we construct the dense core subsets $X_\theta(300, 50)$ in twelve different angle bins $\theta \in \{0, \frac{\pi}{12}, \ldots \frac{11\pi}{12}\}$. For computational feasibility, we then apply sequential maxmin downsampling [20] to reduce each set $X_\theta(300, 50)$ to a subset of 50 data points. Based on Ripser computations we observe that the persistent homology is robust with regard to this downsampling procedure. We then build a zigzag filtration as described above, and use Dionysus [31] to compute the zigzag homology barcodes in Fig. 10. The long interval confirms that the circles indeed piece together compatibly. Furthermore, by checking that the orientation on a generator for the 1-dimensional homology of $X_0(300, 50)$ is preserved after looping once around the circle, we confirm that this fiber bundle structure is that of a torus (instead of a Klein bottle). We remark that another way to verify that this fiber bundle is a torus instead of a Klein bottle would be to use persistence for circle-valued maps [13] (on the map from the total space to the circle that encodes the predominant angle θ of each flow patch).

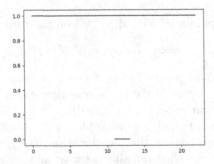

Fig. 10. A 1-dimensional zigzag persistence computation, showing that the circles in Fig. 8 glue together in the structure of a fiber bundle.

We would like to emphasize that the 2-dimensional flow torus model does not model all common optical flow patches, such as zooming in, zooming out, or roll (rotation around the z-axis in Fig. 6 (right)).

6   Conclusions

Using topological machinery, including persistent homology and zigzag persistence, we explore the nonlinear statistics of high-contrast 3×3 optical flow patches from the computer-generated video short *Sintel*. We find that with a global estimate of density, the densest patches lie near a circle. Furthermore, after selecting the optical flow patches whose predominant direction of flow lies in a small bin of angle values, we find that the patches in each such bin are well-modeled by a circle. Combining these bins together provides a torus model for optical flow, which furthermore is naturally equipped with the structure of a fiber bundle over a circular base space of range image patches. As no instrument can measure ground-truth optical flow, an understanding of the nonlinear statistics of flow is needed in order to serve as a prior for optical flow estimation algorithms.

Acknowledgements. We would like to thank Gunnar Carlsson, Bradley Nelson, Jose Perea, and Guillermo Sapiro for helpful conversations.

References

1. Adams, H., Atanasov, A., Carlsson, G.: Nudged elastic band in topological data analysis. Topological Methods Nonlinear Anal. **45**(1), 247–272 (2015)
2. Adams, H., Carlsson, G.: On the nonlinear statistics of range image patches. SIAM J. Imaging Sci. **2**(1), 110–117 (2009)
3. Adams, H., et al.: Persistence images: a vector representation of persistent homology. J. Mach. Learn. Res. **18**(8), 1–35 (2017)
4. Armstrong, M.A.: Basic Topology. Springer, Heidelberg (2013). https://doi.org/10.1007/978-1-4757-1793-8
5. Baker, S., Scharstein, D., Lewis, J., Roth, S., Black, M.J., Szeliski, R.: A database and evaluation methodology for optical flow. Int. J. Comput. Vis. **92**(1), 1–31 (2011)
6. Bao, W., Li, H., Li, N., Jiang, W.: A liveness detection method for face recognition based on optical flow field. In: 2009 International Conference on Image Analysis and Signal Processing, IASP 2009, pp. 233–236. IEEE (2009)
7. Barron, J.L., Fleet, D.J., Beauchemin, S.S.: Performance of optical flow techniques. Int. J. Comput. Vis. **12**(1), 43–77 (1994)
8. Baryshnikov, Y., Ghrist, R.: Target enumeration via euler characteristic integrals. SIAM J. Appl. Math. **70**(3), 825–844 (2009)
9. Bauer, U.: Ripser: a lean C++ code for the computation of Vietoris-Rips persistence barcodes. Software (2017). https://github.com/Ripser/ripser
10. Beauchemin, S.S., Barron, J.L.: The computation of optical flow. ACM Comput. Surv. (CSUR) **27**(3), 433–466 (1995)
11. Bendich, P., Marron, J.S., Miller, E., Pieloch, A., Skwerer, S.: Persistent homology analysis of brain artery trees. Ann. Appl. Stat. **10**(1), 198 (2016)
12. Bubenik, P.: Statistical topological data analysis using persistence landscapes. J. Mach. Learn. Res. **16**(1), 77–102 (2015)
13. Burghelea, D., Dey, T.K.: Topological persistence for circle-valued maps. Discrete Comput. Geom. **50**(1), 69–98 (2013)

14. Butler, D.J., Wulff, J., Stanley, G.B., Black, M.J.: A naturalistic open source movie for optical flow evaluation. In: Fitzgibbon, A., Lazebnik, S., Perona, P., Sato, Y., Schmid, C. (eds.) ECCV 2012. LNCS, vol. 7577, pp. 611–625. Springer, Heidelberg (2012). https://doi.org/10.1007/978-3-642-33783-3_44

15. Carlsson, G.: Topology and data. Bull. Am. Math. Soc. **46**(2), 255–308 (2009)

16. Carlsson, G., De Silva, V., Morozov, D.: Zigzag persistent homologyand real-valued functions. In: Proceedings of the Twenty-Fifth annual Symposium on Computational Geometry, pp. 247–256. ACM (2009)

17. Carlsson, G., Ishkhanov, T., De Silva, V., Zomorodian, A.: On the local behavior of spaces of natural images. Int. J. Comput. Vis. **76**(1), 1–12 (2008)

18. Carlsson, G., de Silva, V.: Zigzag persistence. Found. Comput. Math. **10**(4), 367–405 (2010)

19. Chung, M.K., Bubenik, P., Kim, P.T.: Persistence diagrams of cortical surface data. In: Prince, J.L., Pham, D.L., Myers, K.J. (eds.) IPMI 2009. LNCS, vol. 5636, pp. 386–397. Springer, Heidelberg (2009). https://doi.org/10.1007/978-3-642-02498-6_32

20. De Silva, V., Carlsson, G.: Topological estimation using witness complexes. SPBG **4**, 157–166 (2004)

21. Edelsbrunner, H., Harer, J.L.: Computational Topology: An Introduction. American Mathematical Society, Providence (2010)

22. Edelsbrunner, H., Letscher, D., Zomorodian, A.: Topological persistence and simplification. In: 2000 Proceedings of 41st Annual Symposium on Foundations of Computer Science, pp. 454–463. IEEE (2000)

23. Fleet, D., Weiss, Y.: Optical flow estimation. In: Paragios, N., Chen, Y., Faugeras, O. (eds.) Handbook of Mathematical Models in Computer Vision, pp. 237–257. Springer, Boston (2006). https://doi.org/10.1007/0-387-28831-7_15

24. Geiger, A., Lenz, P., Stiller, C., Urtasun, R.: Vision meets robotics: the KITTI dataset. Int. J. Robot. Res. (IJRR) **32**, 1231–1237 (2013)

25. Hatcher, A.: Algebraic Topology. Cambridge University Press, Cambridge (2002)

26. Horn, B.K., Schunck, B.G.: Determining optical flow. Artif. Intell. **17**(1–3), 185–203 (1981)

27. Huang, J., Lee, A.B., Mumford, D.B.: Statistics of range images. In: Proceedings of the IEEE Conference on Computer Vision and Pattern Recognition, pp. 324–332 (2000)

28. Lee, A.B., Pedersen, K.S., Mumford, D.: The nonlinear statistics of high-contrast patches in natural images. Int. J. Comput. Vis. **54**(1–3), 83–103 (2003)

29. Lum, P., et al.: Extracting insights from the shape of complex data using topology. Sci. Rep. **3**, 1236 (2013)

30. Mac Aodha, O., Humayun, A., Pollefeys, M., Brostow, G.J.: Learning a confidence measure for optical flow. IEEE Trans. Pattern Anal. Mach. Intell. **35**(5), 1107–1120 (2013)

31. Morozov, D.: Dionysus. http://www.mrzv.org/software/dionysus2/

32. Roosendaal, T.: Sintel. Blender Foundation, Durian Open Movie Project (2010). http://www.sintel.org/

33. Roth, S., Black, M.J.: On the spatial statistics of optical flow. Int. J. Comput. Vis. **74**(1), 33–50 (2007)

34. de Silva, V., Ghrist, R.: Coordinate-free coverage in sensor networks with controlled boundaries via homology. Int. J. Robot. Res. **25**(12), 1205–1222 (2006)

35. Topaz, C.M., Ziegelmeier, L., Halverson, T.: Topological data analysis of biological aggregation models. PloS One **10**(5), e0126383 (2015)
36. Xia, K., Wei, G.W.: Persistent homology analysis of protein structure, flexibility, and folding. Int. J. Numer. Methods Biomed. Eng. **30**(8), 814–844 (2014)
37. Zomorodian, A., Carlsson, G.: Computing persistent homology. Discrete Comput. Geom. **33**(2), 249–274 (2005)

Topological Homogeneity for Electron Microscopy Images

Helena Molina-Abril[1(✉)], Fernando Diaz del Rio[1], Maria P. Guerrero-Lebrero[2],
Pedro Real[1], Guillermo Barcena[2], Veronica Braza[3], Elisa Guerrero[2],
David Gonzalez[3], and Pedro L. Galindo[2]

[1] H.T.S. Informatics' Engineering, University of Seville, Seville, Spain
`habril@us.es`
[2] Department of Computer Science and Engineering, University of Cadiz,
11510 Puerto Real, Cadiz, Spain
[3] University Research Institute on Electron Microscopy and Materials (IMEYMAT),
University of Cadiz, 11510 Puerto Real, Cadiz, Spain

Abstract. In this paper, the concept of homogeneity is defined, from a topological perspective, in order to analyze how uniform is the material composition in 2D electron microscopy images. Topological multiresolution parameters are taken into account to obtain better results than classical techniques.

Keywords: Topology · Homogeneity · Electron microscopy · Images

1 Introduction

Microscopy imaging techniques are employed by scientists and researchers to improve their ability to view the microscopic world. The obtained 2D images are used to get information about structure and/or composition distributions of the studied objects and one of the parameters to be usually required is the analysis of the homogeneity. Certainly, the two-dimensional homogeneity question crops up in many different scientific fields. Thus, in quality assurance programs, surface smoothness may be checked in different blocks of material. In biological investigations, the density of a biological tissue may be recorded for purposes of detecting regions of different contrasts [1]. In material science, electron microscopy images are used to evaluate the elemental distribution or strain fields in order to characterize its structure.

In all these situations, the same issue arises. Given a two-dimensional matrix of measurements, it is necessary to assess the randomness [2], the apparition of patterns or microstructure, the lack of gradients, etc. in the image. The formulation of this question overlooks any concept of a formal statistical distribution [5].

This work has been supported by the Spanish research projects MTM2016-81030-P, TEC2012-37868-C04-02 (AEI/FEDER, UE) and the VPPI of the University of Seville.

© Springer Nature Switzerland AG 2019
R. Marfil et al. (Eds.): CTIC 2019, LNCS 11382, pp. 166–178, 2019.
https://doi.org/10.1007/978-3-030-10828-1_13

The question is not how to characterize in a probabilistically way the observations [3,4]; rather it is to characterize how observations could relate with each other locally. In this paper, the concept of homogeneity is defined, from a topological perspective, in order to analyze the homogeneity in 2D electron microscopy images. Topological multiresolution parameters are taken into account to obtain better results than classical techniques.

2 Electron Microscopy Image Data

In recent years, GaAsSbN alloys have been established as an effective material for solar cell applications. To solve the problems that are recurrent in quaternary alloys, a new approach has been adopted by means of the growth of superlattice (SL) structures. These SL structures consists in very thin layers epitaxially grown that could give supplementary advantages over bulk counterparts, such as an extra bandgap tunability via period thickness or better crystal qualities.

In this work, two SL structures have been studied before and after an annealing at 800 °C: type-I (GaAsSbN/GaAs) and type-II (GaAsSb/GaAsN) grown over a GaAs substrate. HAADF and LAADF images were simultaneously acquired in STEM mode in a double aberration corrected FEI Titan3 Cubed Themis operated at 200 kV and used to obtain distribution maps of N from a methodology through the suitable normalization and discrimination of the intensity ratios of HAADF and LAADF images [12]. Figure 1 shows these calculated N mappings where (a) and (b) belong to type-I SL images (before and after RTA process respectively) and (c) and (d) correspond to type-II SL images.

These mappings are built to characterize the distribution of N inside these SL and find its relationship with the photoluminescence and photocurrent properties. Ideally, N should incorporate in a square-wave profile with a constant and homogeneous composition obtained by random distribution of N in V-sites. However, N competes for these lattice positions with As and Sb so there are serious difficulties in the precise control of the Sb and N contents [7,11]. In addition, the extremely low solubility of N in these alloys could favors the formation of N-rich regions [10,13,16]. All of this suggests that the growth of high quality III-V-N structures may be difficult to achieve [7,8,14]. Several approaches has been proposed to overcome these issues. On one hand, it is suggested the spatial separation of Sb and N atoms as in type-II SL could avoid the ubiquitous growth problems during the simultaneous growth of Sb and N that happens in type-I improving crystal quality [9]. On the other hand, it is believed that post-growth annealing processes may also increase the homogeneity of N within the layer by decreasing composition fluctuations. In addition, device performance is a strong function of the quality of the interfaces. The formation of atomically flat interfaces by suppressing surface undulations is fundamental to enhance their optical properties [6,15]. In each case, to check the possible improvements of the crystal quality, it is necessary to compare and evaluate both approaches and the effect of the thermal annealing in terms of homogeneity.

The study of homogeneity, among others, could determine this closeness/distance of the ideal design. Until now, to calculate the homogeneity of

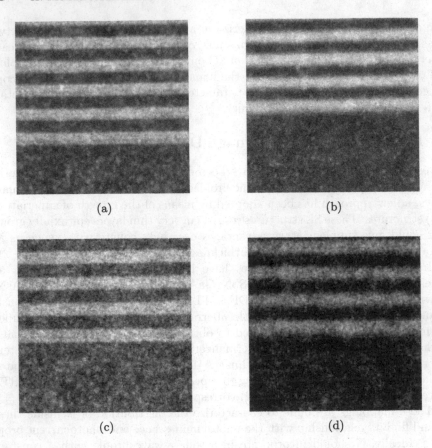

Fig. 1. Mappings of the N distribution of type-I (a) and type-II (c) as grown and after annealing (b) y (d), respectively. We used a grey scale, where white is indicative of the highest N content and black of the lowest one.

these images, the standard deviation of the values has been used to estimate it. The higher the standard deviation, the lower the homogeneity. However, this procedure is too simple for 2D analyses, as they do not take into account the possibility of regular gradients or patterns that may add a certain degree of symmetry or order to the material, the presence or degree of clusters as well as the abruptness or roughness of the interfaces. Considering all this, it seems necessary to use topological techniques to define homogeneity in order to obtain better results.

3   Related Works

Current methods dealing with the problem of measuring how far a given material is from a constant composition, clearly fail when trying to analyze the homogeneity of 2D electron microscopy images.

Linear homogeneity measures have been used in some cases for some simple images. For example, the mean and standard deviation of the gray frequency histogram can give an approximate idea of the two-dimensional homogeneity when the set of images are very different in gray levels. However, for the general case these simple values are not relevant at all. Some plain examples illustrate this issue. Let us consider two 256×256 images with very different "homogeneity aspects". Firstly, a chessboard-like image with interlaced values of 0 and 255, and secondly an image with two parts: one is pure black (0) and the other pure white (255). Neither arithmetic mean nor standard deviation provide distinction at all among these extreme cases (they are 127.500 and 127.501 resp.). We are conscious that the concept of "homogeneity" is being intuitively expressed by now, but the difference between both images obvious. Another almost homogeneous image where mean and deviation return paradoxically high values is an image where each row has the gray level of the precedent plus one (in our case, 256 rows with grey from 0 to 255). In this case, mean results also in 127.500 and deviation in 73.901. Hence, it is clear that simple measures are not valid to express the homogeneity of grey level images.

Here we present a topology-based method, in which standardized topological numbers provide a robust quantity for measuring how uniform is the material composition.

There have been another more sophisticated attempts in using digital topology for measuring image's homogeneity. One of the first one was presented in [19]. However, authors there did not provide a quantity for comparing homogeneity among images, but a multiresolution representation that can be used for texture characterization. More recent methods dealing with texture classification can be seen in [17,20]. These works deal with texture classification by applying machine learning algorithms to a set of features obtained from the image. Other more complicated functionals have also been proposed for comparing noisy experimental image data with statistical models (see [18]). However, to the best of our knowledge, there are no results satisfying that: (1) Have been successfully applied to Electron Microscopy Image data, and (2) provide a normalized homogeneity measure for image comparison.

4   Topological Techniques

Let us first recall basic notions in the field of digital images and digital topology. We denote by $\mathbb{Z}$ the set of relative integers. A point $x \in \mathbb{Z}^2$ is defined by (x_1, x_2) with $x_i \in \mathbb{Z}$. A 2D grayscale image may be seen as a map $\mathcal{I}$ from $\mathbb{Z}^2$ to $\mathbb{Z}$. For each point $x \in \mathbb{Z}^2$, $\mathcal{I}(x)$ is the (graylevel) intensity value of x. A binary 2D image is then seen as the map $\mathcal{B}$ from $\mathbb{Z}^2$ to $\{0, 1\}$.

For each point (pixel) in a given 2D image, we consider two local adjacency relations Γ_4 and Γ_8 defined: For each point $x \in \mathbb{Z}^2$:

$$\Gamma_4(x) = \{y \in \mathbb{Z}^2; |y_1 - x_1| + |y_2 - x_2| \leq 1\}$$

$$\Gamma_8(x) = \{y \in \mathbb{Z}^2; max(|y_1 - x_1|, |y_2 - x_2|) \leq 1\}$$

In the following, we will denote by n a number such that $n = 4$ or $n = 8$. We define $\Gamma_n^*(x) = \Gamma_n(x)\setminus\{x\}$. The point $y \in \mathbb{Z}^2$ is n-adjacent to $x \in \mathbb{Z}^2$ if $y \in \Gamma_n(x)$.

In this section, electron microscopy images will be analyzed from a topological point of view, by first of all, creating their corresponding so called "crack images". Crack images are binary images created by considering each possible gray level difference among adjacent pixels. That is, if the gray-scale of image $\mathcal{I}$ is composed by G gray levels, we will create $G - 1$ binary crack images for image $\mathcal{I}$ (plus an initial gray scale image denoted as $\mathcal{I}_0^{Crk}$). From now on, the number of possible gray levels in a given image $\mathcal{I}$ will be denoted as G, and the different levels of gray that are actually present in $\mathcal{I}$ will be denoted as $g_1, \ldots g_l$. In the following Figures with binary images, value zero will be represented in black color, and value one in white color.

First of all, if $\mathcal{I}$ is a ($Nrows \times Ncols$) image, a new gray-scale image $\mathcal{I}_0^{Crk}$ with (($2*Nrows+1) \times (2*Ncols+1)$) pixels is created by adding a black frame of width 1 surrounding the original image, and adding a new pixel between each pair of 4-adjacent pixels in $\mathcal{I}$. All these added pixels are set to the following values: Pixels belonging to the frame surrounding the original image are set to zero, and will be denoted as $\mathcal{F}$. Pixels added between each pixel in $\mathcal{I}$ are set to one. These pixels in $\mathcal{I}_0^{Crk}$ will be denoted as $\mathcal{A}$. All the other pixels (the ones corresponding to the original image) are set to their corresponding value in $\mathcal{I}$ and will be denoted as $\mathcal{O}$. Figure 2(a) shows a synthetic image whose gray-scale contains 6 levels that are all of them present in the image ($G = 6$, $l = 6$, $g_1 = 0$ and $g_6 = 5$). An example of its corresponding $\mathcal{I}_0^{Crk}$ image is depicted in Fig. 2(b), where $\mathcal{F}$ are set to zero, $\mathcal{A}$ are set to one and $\mathcal{O}$ are set to their original values. Then, for the construction of the corresponding crack images, the gray level will be used as a threshold for checking intensity differences among each 4-adjacent pixel in the original image.

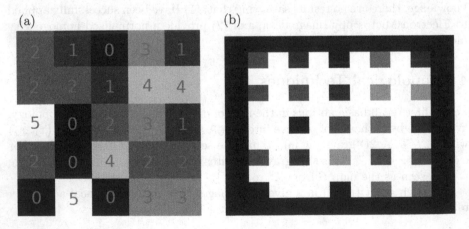

(a)   (b)

Fig. 2. (a) Synthetic image $\mathcal{I}$ showing different gray intensities and (b) its corresponding $\mathcal{I}_0^{Crk}$

Algorithm 1 shows the process of crack images generation. The main idea here, is that for each possible gray-value difference d, the corresponding crack image $\mathcal{I}_d^{Crk}$ represents fissures separating image regions whose gray-level difference is above (or equal to) this quantity d.

Algorithm 1. [Creating crack images]

Input: A digital 2D Image $\mathcal{I}$ in a gray-scale of G values

    Generate $\mathcal{I}_0^{Crk}$, and define pixels in $\mathcal{A}$ and pixels in $\mathcal{O}$

    for $d = 1$ to G **do**

        for each point $x \in \mathcal{A}$ **do**

            if $\exists\ y, z \in \Gamma_4^*(x) : y, z \in \mathcal{O}$ and $|\mathcal{I}_0^{Crk}(y) - \mathcal{I}_0^{Crk}(z)| >= d$ **then**

                $\mathcal{I}_d^C(x) = 0$

        for each point $x \in \mathcal{A}$ **do**

            if $\exists\ y, z \in \Gamma_4^*(x) : y, z \in \mathcal{A}, \mathcal{I}_d^{Crk}(y) = 0$ and $\mathcal{I}_d^{Crk}(z) = 0$ **then**

                $\mathcal{I}_d^{Crk}(x) = 0$

        for each point $x \in \mathcal{O}$ **do**

            $n_0 = \#\{y \in \Gamma_8^*(x) : \mathcal{I}_d^{Crk}(y) = 0\}$

            $n_1 = \#\{y \in \Gamma_8^*(x) : \mathcal{I}_d^{Crk}(y) = 1\}$

            if $n_0 > n_1$ **then**

                $\mathcal{I}_d^C(x) = 1$

            else

                $\mathcal{I}_d^C(x) = 0$

Output: A set of G crack images $\mathcal{I}_d^{Crk}$

The corresponding crack images for the synthetic image in Fig. 2(a) are shown in Fig. 3(a) to (e) for $d = 1$, $d = 2$, $d = 3$, $d = 4$ and $d = 5$ respectively. Note that at the beginning of the algorithm, all the pixels inserted between two pixels of the original image (denoted as $\mathcal{A}$) are set to one in $\mathcal{I}_0^{Crk}$. Taking for instance $d = 1$, and following Algorithm 1 the intensity difference among pixels is almost always greater or equal to one, so most of the added pixels will change their value to zero, giving as a result the image shown in Fig. 3(e) in which only six of these added pixels remain unchanged (value equal to one, colored in white).

Once a crack image is created for each gray-level difference, the number of connected black and connected white components are computed for each one of them. These numbers, corresponding to the well known Betti numbers of dimension 0 and 1 respectively, will be used a signature of the original image homogeneity. More exactly, the normalized sum of these Betti numbers along the gray spectrum is going to be demonstrated as a very robust measure of real image homogeneity. From now on, these sums will be referred here as B_0 (for the black components) and B_1 (for the white components).

The computation of a normalized sum of the well-known Euler number along the gray spectrum, i.e. $B_1 - B_0$ in our notation, is shown in Algorithm 2. The resulting value will be the homogeneity index for the analyzed images. Figure 3(f), shows the evolution of β_0 and β_1 with the progress of Algorithm 2

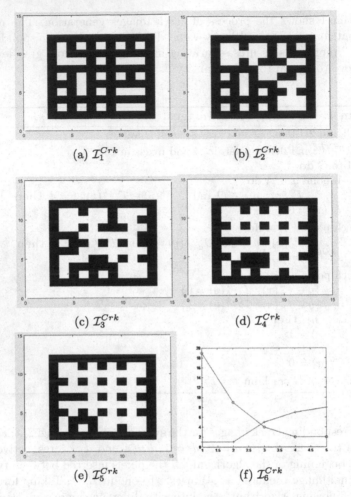

Fig. 3. Crack images corresponding to image in Fig. 2(a) and the evolution of homogeneity measures along all the possible gray differences. The line with crosses represents the β_0 whereas line with circles the β_1

(with d going from 1 to 5) for Image in Fig. 2(a). The evolution of β_0 is represented with crosses, and the evolution of β_1 with circles in the image.

According to previous construction of the crack images and homogeneity measures, it is interesting to observe the behavior of extremal (homogeneous vs. heterogeneous) images. Those images have the biggest and smallest $B_1 - B_0$ values respectively. These values, will be used for normalizing our B_0 and B_1 measures, so at the end, $-1 \leq B_1 - B_0 \leq 1$. On one hand, given a pure homogeneous image (of any gray constant level), all its crack images are the same (differing in size) as the one shown in Fig. 4(a). Analyzing β_0 and β_1 in these crack images, we will obtain the maximum number of connected black

---

Algorithm 2. [Computing the homogeneity measure]

Input: A set of $\mathcal{I}_d^{Crk}$ Crack Images
   $B_0 = 0, B_1 = 0$
   for $d = 1$ to G **do**
      $\beta_0 =$ Betti number of dimension 0 of $\mathcal{I}_d^{Crk}$
      $\beta_1 =$ Betti number of dimension 1 of $\mathcal{I}_d^{Crk}$
      $B_0 = B_0 + \beta_0$
      $B_1 = B_1 + \beta_1$
   $B_0 = Normalize(B_0)$
   $B_1 = Normalize(B_1)$
   Output: $B_1 - B_0$

---

components β_0 and only one connected white component β_1. Note that the surrounding black frame inserts a black component that "touches" all the components in the four image borders, thus the number of maximum connected black components in the crack images is $1 + (Nrows - 2) * (Ncols - 2)$, and the number of maximum connected white components is $(Nrows) * (Ncols)$, where $Nrows$ and $Ncols$ are the number of rows and columns of the original image. In the same way, the minimum number of connected black components and connected white components is one. Therefore, the crack image shown in Fig. 4(a) corresponding to a 5×5 homogeneous image, has $\beta_0 = 10$ and $\beta_1 = 1$.

On the other hand, Fig. 4(b) depicts the crack image for the maximum possible heterogeneous 5×5 image: a chessboard-like image with interlaced gray values of g_1 and g_l. All the possible crack images for gray-level differences in the range $[1, (g_l - g_1)]$ result in the same form (that of Fig. 4(b)), thus having the maximum number of white components (5×5 in this case) but only one black component. Therefore, the crack image shown in Fig. 4(b) has $\beta_0 = 1$ and $\beta_1 = 25$.

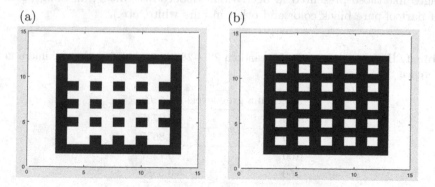

(a)   (b)

Fig. 4. (a) Crack image corresponding a homogeneous image. (b) Crack image corresponding to a heterogeneous chess-like image.

Thus, summing up β_0 and β_1 for any possible gray difference, in the case of an homogeneous image, B_1 has the minimum value $(G-1)$ (the result of summing up $G-1$ times the value one for $\beta_1 = 1$), and B_0 has the maximum value $(G*(1+(Nrows-2)*(Ncols-2)))$ where G is the number of possible gray-levels in the original image. Therefore, the normalization of our homogeneity measure B_1-B_0 will be commputed by dividing B_0 by $(G*(1+(Nrows-2)*(Ncols-2)))$ and B_1 by $(G*((Nrows)*(Ncols)))$. Doing that, we obtain $B_1 - B_0 = 1$ for a purely heterogeneous image (a chessboard-like image with interlaced extreme gray values) and $B_1 - B_0 = -1$ for a purely homogeneous image.

5   Experimentation

The relation between parameter $B_1 - B_0$ and homogeneity can be demonstrated by analyzing random synthetic images with different maximum and minimum gray levels. It is expected that those images with bigger gray contrasts between their pixels will be more heterogeneous, whereas the images with similar gray levels will present a high degree of homogeneity. Besides it is also required for a good homogeneity parameter that the size of the image has no influence on its value.

Table 1 shows the homogeneity results for different random gray-scale 28×28 images in which the maximum grey level has been modified. Figure 5 shows the evolution of the number of holes and connected components when the contrast is increased for the crack images of two synthetic random images. As stated in previous section, the more contrast the crack image has, the more connected components appear (that is, bigger β_0) and the less holes are found (i.e., smaller β_1). It is worth to mention that the size of the image has a negligible influence on the homogeneity values (slight variations because of the random generation of the images). Therefore, we can conclude that the bigger $B_1 - B_0$ is, the more heterogeneous the image is. No doubt that this analysis is far obvious for simple images like those presented in Sect. 3 (chessboard-like image, an image with a half part of pure black color and other in pure white, etc.).

Table 1. Homogeneity results for random 28×28 images varying the maximum gray difference

Maximum grey level	$B_1 - B_0$
10	−0.9735
50	−0.8645
100	−0.7257
150	−0.5883
200	−0.4655
255	−0.2908

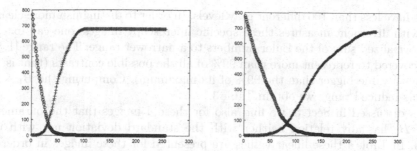

Fig. 5. (a) Representation of the number of holes and connected components vs. grey level for the crack images of two synthetic random images with maximum gray intensities of 100 (Left) and 255 (Right). The line with crosses represents the β_0 whereas line with circles the β_1

Experimentation has been carried out using images described in Sect. 2. Four of these images are shown in Fig. 1. Three samples of 100×100 pixels belonging to the white bands have been taken for each image (see Fig. 6 where samples (a) correspond to Fig. 1(a), samples (b) to Fig. 1(b), etc.).

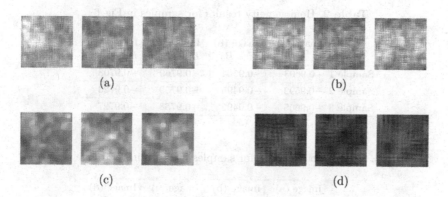

Fig. 6. Electron microscopy samples corresponding to images in Fig. 1

The resulting homogeneity measure $B_1 - B_0$ is shown in Table 2. As we can see, $B_1 - B_0$ values are very similar for any fragment extracted for the same image (around -0.960 for image (a), around -0.950 for image (b), around -0.976 for image (c) and around -0.972 for image d). More exactly, means for the three segments of images (a), (b), (c) and (d) are -0.9600, -0.9497, -0.9762, -0.9719, whereas standard deviation results to be 0.0007, 0.0006, 0.0006, 0.0010. This clearly points out the robustness of the proposed homogeneity measure $B_1 - B_0$.

Besides, the resemblance of all $B_1 - B_0$ values and their proximity to -1.0 indicate that the white bands are very homogeneous for all the tested electronic microscopy images. In fact, image histograms reveal that more than 95% of the pixels have less than 100 different gray levels. In order to distinguish more clearly and with divergent measures these specific microscopy images, one can extend the normalized sum of the Euler numbers to a narrower range. The range $[1, 50]$ is discovered to represent more than 99% of all the possible contrasts (that is, β_0 reaches a value bigger than the 99% of its maximum). Computing the $B_1 - B_0$ for this reduced range we obtain Table 3.

As discussed in Sect. 3, we find also for these 4 images that the parameter $B_1 - B_0$ has any relation neither with the standard deviation nor with the mean. In Table 4 these linear values are presented for these images in order to corroborate that they would arrange the images in a completely different order than $B_1 - B_0$ would.

Finally, we can conclude for the microscopy images that they can be arranged with respect to their homogeneity in this order (from smaller to bigger): b, a, d, c. This parameter is currently helping microscopy imaging experts to analyze how thermal processes affect the structure and/or composition of different material.

In future works we expect that the homogeneity parameter described here, being a well defined and robust measure, would serve to analyze other images where texture and regularity play an important role, like those of skin cancer, granulometry, porosity materials, etc.

Table 2. Homogeneity results for samples in Fig. 6

	Image (a) $B_1 - B_0$	Image (b) $B_1 - B_0$	Image (c) $B_1 - B_0$	Image (d) $B_1 - B_0$
Sample 1	−0.9603	−0.9504	−0.9769	−0.9708
Sample 2	−0.9592	−0.9495	−0.9759	−0.9723
Sample 3	−0.9605	−0.9492	−0.9758	−0.9727

Table 3. Homogeneity results for samples in Fig. 6 in the range $[1, 50]$

	Image (a) $B_1 - B_0$	Image (b) $B_1 - B_0$	Image (c) $B_1 - B_0$	Image (d) $B_1 - B_0$
Sample 1	−0.7651	−0.7172	−0.8467	−0.8166
Sample 2	−0.7598	−0.7128	−0.8416	−0.8242
Sample 3	−0.7665	−0.7113	−0.8413	−0.8261

Table 4. Standard deviation and mean for samples in Fig. 6

	Image (a)	Image (b)	Image (c)	Image (d)
Standard deviation	45.0804	45.1256	37.4100	27.7506
Mean	194.4156	189.1598	159.2374	92.0668

6   Conclusions

In this paper, the concept of homogeneity is defined, from a topological perspective, in order to analyze the homogeneity in 2D electron microscopy images. A standardized topological number is provided as a robust quantity for measuring how uniform is the material composition.

The proposed topological number has been applied to numerically analyze the homogeneity of nitrogen distribution in composition maps obtained from ADF STEM images in type I (GaAsSbN/GaAs) and type II (GaAsSb/GaAsN) superlattice structures before and after RTA. A range of homogeneity between the samples has been established. First, it is shown that the SL-I image gives a parameter $B_1 - B_0$ higher than the SL-II. Secondly, our calculations show that annealing treatment results in a significant increase in uniformity in both types of SLs, i.e. better diffusion leads to a homogenization of the distribution of N within the layer. Our analysis has revealed that this increase during RTA is higher in SL-II than in SL-I and this result is in agreement with the photoluminescence results.

Further work extending this measure to higher dimensions and considering other more sophisticated topological relations will be performed in the future. Extensions to other possible applications in which homogeneity measures might be useful will be assessed as well (biomedical, astronomical images, etc.).

References

1. Miller, R.L., Kahn, J.S.: Statistical Analysis in the Geological Sciences. Wiley, New York (1962)
2. Yaglom, A.M.: An Introduction to the Theory of Stationary Random Functions. Prentice Hall, Englewood Cliffs (1962)
3. Materm, B.: Spatial variation. Comm. Swed. For. Res. Inst. **49**, 144 (1960)
4. Ripley, B.: Spatial Statistics. Wiley, New York (1981)
5. Cramer, H., Leadbetter, M.R.: Stationary and Related Stochastic Processes. Wiley, New York (1967)
6. Ahsan, N., Miyashita, N., Islam, M., Yu, K., Walukiewicz, W., Okada, Y.: Effect of Sb on GaNAs intermediate band solar cells. IEEE J. Photovoltaics **3**(2), 730–736 (2013)
7. Braza, V., et al.: Sb and N incorporation interplay in GaAsSbN/GaAs Epilayers near lattice-matching condition for 1.0–1.16-eV photonic applications. Nanoscale Res. Lett. **12**(1), 356 (2017)
8. Cheah, W.K., Fan, W.J., Wicaksono, S., Yoon, S.F., Tan, K.H.: Low antimony-doped GaNxAs1-x on GaAs grown by solid-source molecular-beam epitaxy. J. Cryst. Growth **254**(3–4), 305–309 (2003)
9. Gonzalo, A., et al.: Strain-balanced type-II superlattices for efficient multi-junction solar cells. Sci. Rep. **7**(1), 4012 (2017)
10. Ho, I.H., Stringfellow, G.B.: Solubility of nitrogen in binary III–V systems. J. Cryst. Growth **178**(1–2), 1–7 (1997)
11. Reyes, D.F., et al.: Modelling of the Sb and N distribution in type II GaAsSb/GaAsN superlattices for solar cell applications. Appl. Surf. Sci. **442**, 664–672 (2018)

12. Ruiz-Marin, N., et al.: Nitrogen mapping from (HA) ADF analysis in quaternary dilute nitride superlattices 1 introduction. Appl. Surf. Sci. (in Review)

13. Stringfellow, G.B.: Thermodynamic considerations for epitaxial growth of III/V alloys. J. Cryst. Growth **468**, 11–16 (2017)

14. Wu, L.J., et al.: MBE growth and properties of GaAsSbN/GaAs single quantum wells. J. Cryst. Growth **279**(3–4), 293–302 (2005)

15. Wu, Z.H., et al.: Spontaneous formation of highly regular superlattice structure in InGaN epilayers grown by molecular beam epitaxy. Appl. Phys. Lett. **98**(14), 4–7 (2011)

16. Zhang, S.B., Wei, S.H.: Nitrogen solubility and induced defect complexes in epitaxial GaAs:N. Phys. Rev. Lett. **86**(9), 1789–1792 (2001)

17. Barros Neiva, M., Vacavant, A., Martinez Bruno, O.: Improving texture extraction and classification using smoothed morphological operators. Digit. Sig. Process. **83**, 24–34 (2018)

18. Mantz, H., Jacobs, K., Mecke, K.: Utilizing Minkowski functionals for image analysis: a marching square algorithm. J. Stat. Mech. Theory Exp. **12**, 12015 (2008)

19. Pikazi, A., Averbuch, A.: An efficient topological characterization of gray-levels textures, using a multiresolution representation. Graph. Models Image Process. **59**(1), 1–17 (1997)

20. Sonali Dash, S., Ranjan Jena, U.: Multi-resolution Laws' Masks based texture classification. J. Appl. Res. Technol. **15**, 571–582 (2018)

Characterising Epithelial Tissues Using Persistent Entropy

N. Atienza[1], L. M. Escudero[2,3], M. J. Jimenez[1(✉)], and M. Soriano-Trigueros[1]

[1] Departamento Matematica Aplicada I, Universidad de Sevilla,
Campus Reina Mercedes, 41012 Sevilla, Spain
{natienza,majiro,msoriano4}@us.es
[2] Departamento de Biología Celular, Universidad de Sevilla, Sevilla, Spain
[3] Instituto de Biomedicina de Sevilla (IBiS), Hospital Universitario Virgen del Rocio,
CSIC, Universidad de Sevilla, 41013 Sevilla, Spain
lmescudero-ibis@us.es

Abstract. In this paper, we apply persistent entropy, a novel topological statistic, for characterization of images of epithelial tissues. We have found out that persistent entropy is able to summarize topological and geometric information encoded by α-complexes and persistent homology. After using some statistical tests, we can guarantee the existence of significant differences in the studied tissues.

Keywords: Topological data analysis · Persistent entropy Epithelial tissues

1 Introduction

Topological Data Analysis (TDA), originally, had its main motivation in the study of topological analysis of point cloud data. Nowadays is becoming a powerful tool for the study of shape of data, in its most general meaning. The main tool used in TDA is *persistent homology* [4,16], which studies the evolution of homology classes and their life-times (persistence) in an increasing nested sequence of spaces (that is called a filtration) and which is more informative that the homology class of the whole space.

Persistent homology has proved to be a useful tool in the study of shape analysis (in [9], some trends are described). Lately, most of the efforts in the area have been focused on developing a vector representation (obtained via persistent homology) that can be treated, afterwards, from machine learning point of view (see, for example, [1]). Such approaches have got the drawback of the need of large sample datasets, which is not usually the case in practice. For that reason,

Partially supported by MINECO, FEDER/UE under grant MTM2015-67072-P. Authors names are listed in alphabetical order.

we are concerned with exploring the power of persistent entropy, which is a simple parameter that can be obtained from persistent homology and can be statistically studied.

Our main contribution is the application of persistent entropy as a tool for the characterization of biological tissues.

In the following Section, we describe the biological problem that motivated this work. Section 3 recalls main concepts from TDA that will be used in the sequel. Section 4 describes the particular way in which we make use of persistent homology concepts to topologically analyze the input data. Reports on the computations performed as well as some conclusions are collected in Sect. 5. We draw some ideas for future work in the last section.

2   Motivation

Epithelial cells are cells from animal embryos that will transform into one of the 4 types of adult tissues: epithelia, connective tissue, nervous tissue and muscle tissue. Epithelia are packed tissues formed by tightly assembled cells. Their apical surfaces are similar to convex polygons forming a natural tessellation. Epithelial organization has been analyzed in various systems from a topological and biophysical perspective [8,10,13,14]. These studies have been mainly based in the analysis of the polygon distribution of the tissues. A new approach has just been developed in [15], were the authors have provided an image analysis tool (implemented in the open-access platform FIJI) to quantify epithelial organization based in computational geometry and graph theory concepts.

Another work that has to be mentioned is [12], where the authors make use of some image processing tools, like morphological gradient, to classify coating epithelial cells into three different types.

In [11], the authors first applied persistent homology, looking for other organizational traits that could improve the characterization of epithelia. Some initial experiments were described, working on two types of tissues: chick neuroepithelium (cNT) from chicken embryos and wing imaginal disc in the prepupal stage (dWP) from Drosophila. However, we would like also to compare the latter (dWP) with middle third instar wing discs (dWL), which are two proliferative stages separated by 24 h development (and hence, with very similar organization). In this paper, we are concerned with the study of the discriminative ability of persistent entropy, discovering statistically significant differences between images of the three tissues. This work may open a door to the inclusion of persistent entropy as one more parameter to be taken into account in analysis tools like [15].

3   Background

The input used more frequently in topological data analysis is a point cloud in a metric space. In particular, we will work with points in the Euclidean space $\mathbb{R}^2$ obtained from images. The procedure when applying persistent homology is the following. First, transform the information carried by the point cloud into a sequence of geometric figures called a filtration of simplicial complexes. Then, compute the homology (which intuitively can be seen as "holes") for each simplicial complex and track how it evolves along the filtration. Finally, use a suitable way of representing the output and apply statistical methods to reach the conclusions.

In this section we will define briefly these concepts. For a more concise introduction the reader could refer to [5].

Simplicial Complex. A *simplex* is the convex hull of a finite set of linearly independent points $\tau = \{p_1, \ldots, p_n\}$. Any of the possible simplices contained in τ are called its *faces*. A *simplicial complex* $\mathcal{K}$ is formed by a set of simplices satisfying:

1. Every face of a simplex in $\mathcal{K}$ is also in $\mathcal{K}$.
2. The intersection of two simplices in $\mathcal{K}$ is a face of both.

Filtration. A filtration is a finite increasing sequence of simplicial complexes

$$\mathcal{K}_1 \subset \mathcal{K}_2 \ldots \subset \mathcal{K}_n = \mathcal{K}$$

It is commonly defined using a monotonic function $f : \mathcal{K} \to \mathbb{R}$ by which we mean that for $\delta, \tau \in \mathcal{K}$, $f(\delta) \leq f(\tau)$ if $\delta \subset \tau$. In this way, if $a_1 \leq \ldots \leq a_n$ are the function values of the simplices in $\mathcal{K}$, then $\mathcal{K}_i = f^{-1}(-\infty, a_i]$.

Persistent Homology. The inclusion $\mathcal{K}_i \subset \mathcal{K}_{i+1}$ induces a linear map between vector spaces $H_*(K_i) \to H_n(K_{i+1})$, where H_n is the homology of dimension n. Intuitively when a homology class disappears (i.e., it is in K_i but not in K_{i+1} for some i), we say that it dies at time i. When a homology class appears by the first time (i.e., it is in K_i but not in K_{i-1} for some i), we say that it has been born at time i.

Barcodes. The fact that an homology class is born at time i and dies at j can be represented by an interval (i, j). Then, the output of persistent homology can be represented as a multiset $\{(i, j)\}$, where (i, j) are birth-death values of arising homology classes. This is usually represented using barcodes as in Fig. 1.

Persistent Entropy. Persistence barcodes represent reliably the persistent homology output that could be treated statistically. However, the statistical tools used should be stable and robust to noise in the input data. Under this premise, we make use of *persistent entropy* [3], which can be defined as an adaptation of Shannon entropy to this context.

Represent the multiset of birth and death using pairs $\{(x_i, y_i)\}_{i=1}^n$ (note that there might be repeated pairs). Consider the length of each of them $\ell_i = y_i - x_i$ and the total length $L = \ell_1 + \ldots + \ell_n$. Then, the persistent entropy of a barcode is the value

$$E = -\sum_{i=1}^n \frac{\ell_i}{L} \log(\frac{\ell_i}{L}).$$

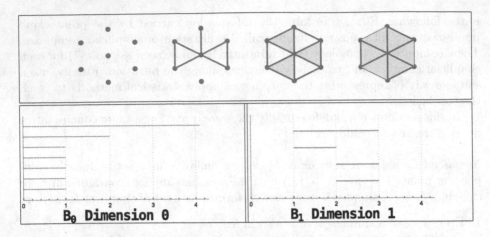

Fig. 1. Top: example of a filtration $\mathcal{K}$. Bottom: barcodes representing connected components (0-th dimensional Betti numbers) and cycles (1-th dimensional).

The maximum possible value of persistent entropy is $\log(n)$ and is reached when all intervals have the same length. The minimum value is 0 and coincides with the case $n = 1$. In general, the greater the number of intervals is and the more homogeneous they are, the greater the persistent entropy is.

The following result guarantees that persistent entropy is robust to small perturbations in the input data (the proof appears in [2]). Before, we need some notation remarks: consider two barcodes $\{B_j\}_{j=1,2}$, given by $B_j = \{(x_i^j, y_i^j)\}_{i=1\ldots n_j}$. The lengths of the bars are $\ell_i^j = y_i^j - x_i^j$ and their total length $L^j = \ell_1^j + \ldots \ell_{n_j}^j$. We denote the bottleneck distance for barcodes as d_∞.

Theorem 1. *Let $\mathcal{K}$ be a simplicial complex and let $f_1, f_2 : \mathcal{K} \to \mathbb{R}$ be two monotonic functions, B_1, B_2 their corresponding barcodes and $n_{max} = \max\{n_1, n_2\}$. Then, if $d_\infty(B_1, B_2) \leq \frac{1}{8} \frac{\max\{L^1, L^2\}}{n_{\max}}$ and $\|f_1 - f_2\| \leq \delta$, we have:*

$$|E(B_1) - E(B_2)| \leq \frac{4\delta n_{\max}}{\max\{L^1, L^2\}} \left[\log(n_{\max}) - \log\left(\frac{4\delta n_{\max}}{\max\{L^1, L^2\}} \right) \right].$$

In other words, this theorem implies that if a maximum number of bars and a minimum length are fixed, then persistent entropy is uniformly continuous respect to the maximum norm of filter functions.

4   Methodology

In this section we will explain the steps involved in the method developed in this paper:

1. Normalize each image so that they all have the same number of cells.
2. Consider the point cloud given by the centroids of the cells.

3. Construct a simplicial complex called Delaunay Triangulation and a filtration on it, called the α-complex, from the point cloud.
4. Compute its persistent homology and persistent entropy.
5. Perform a statistical study and analyze the results.

The input is an image with 1024×1024 pixels. This image is a gray scale image in which each segmented region corresponding to a cell has been labeled with an ID number and pixels on the boundary of cells are labeled by 0. In any case, the images from the database were revised manually to remove small noise and abnormal situations like disconnected cells. In fact, in practice, boundaries of regions are made up by a (at least) 2-pixels thick edge of 4-connected pixels what ensures a good separation of connected components representing each cell.

Now, we further develop some steps in the process:

As for Step 1, an important drawback when using persistent entropy is that the number of cells affects its value. Then, if we want to measure topological features using this parameter, we need to have the same number of cells for each sample. This way, under the assumption that cells have a homogeneous distribution across the images, we have designed the next algorithm:

Input: $n \in \mathbb{N}$ and $M \in (\mathbb{N}^0)^{1024 \times 1024}$.
Output: A set $\mathcal{C}$ of n cells.
$\mathcal{C} := \emptyset$
$x = y = 512$
if $M(x,y) \neq 0$
    $\mathcal{C} := \{M(x,y)\}$
i = 0
while $\#\mathcal{C} < n$
    i = i+1
    Repeat i times
        if $\#\mathcal{C} < n$
            $y := y + (-1)^{i+1}$
            if $M(x,y) \neq 0$ and $M(x,y) \notin \mathcal{C}$
                then $\mathcal{C} := \mathcal{C} \cup \{M(x,y)\}$
    Repeat i times
        if $\#\mathcal{C} < n$
            $x := x + (-1)^i$
            if $M(x,y) \neq 0$ and $M(x,y) \notin \mathcal{C}$
                then $\mathcal{C} = \mathcal{C} \cup \{M(x,y)\}$

The starting point and the orientation do not pursuit any special goal. Their choice has been arbitrary and has been taken always the same for all the images. This fact does not affect the independence of samples. Figure 2 (bottom) shows a simple example in which, taking as input $n = 6$ and the depicted pixel values, the output set of cells would be $\mathcal{C} = \{4, 3, 5, 8, 1, 7\}$.

Centroids of the cells are obtained using the Matlab command *regionprops*. Regarding Step 3, the main difficulty of this methodology is finding the proper filtration to distinguish the cell tissues: α-complex represents a good approximation for cells which are close to be convex. Therefore, this study will focus

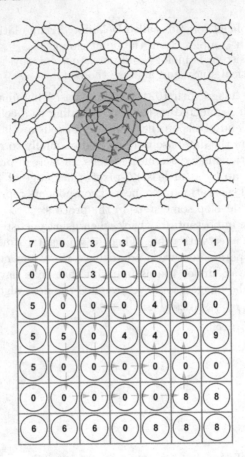

Fig. 2. Top picture illustrates the intuition behind the algorithm to restrict to a proper number of cells. Bottom: flow of the process at pixel level. Each pixel value is the label corresponding to each segmented cell, while label 0 means boundary pixels.

in epithelial cells with this property. Nevertheless, we plan to design in future a new and more general filtration, as mentioned in Sect. 6. Here, we recall the basic concepts involved in the construction of an α-complex:

Voronoi Diagram. A Voronoi Diagram is a partitioning of the plane depending on a set of vertices. For each vertex v_i and $x \in \mathbb{R}^2$ we define the function $f_i(x) = d(v_i, x)$ and a region given by

$$V_i = \{x \mid f_i(x) \leq f_j(x) \quad \forall j\}.$$

We will define a filtration in the following simplicial complex: consider a finite set of points; an edge joins two vertices if the intersection of their Voronoi regions is not empty; the 2-simplices are formed when three points have all possible edges between them. When the points are in general position, there are no simplex with greater dimension and it is called a Delaunay triangulation. Nevertheless, as we are only interested in dimension 0 and 1 we do not need general position assumption and just ignore the higher dimensional simplices.

Alpha Complex. Define B_r^i as the ball of center u_i and radius r. For each r, consider the region $U_r^i = B_r^i \cap V_i$ and define the simplicial complex $\mathcal{K}_r$ with simplices

$$\tau = [u_0 \ldots u_k] \in \mathcal{K}_r \Leftrightarrow U_r^i \cap U_r^j \neq \emptyset \quad i,j = 0 \ldots k.$$

This filtration is called α-complex. See Fig. 3 for a picture.

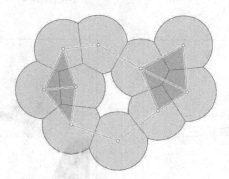

Fig. 3. Example of an Alpha Complex for a fixed r appearing in [5].

In our context, the starting points will be the centroids of the cells. After computing their α-complex, we compute persistent entropy from persistence bar-codes and make a statistical analysis of the results, what will be detailed in the next section.

5 Experiments and Results

Our database consists on 16 images of chick neuroepithelium (cNT), 15 images of Drosophila wing imaginal disc from the third instar larva (dWL) and 13 from the prepual state (dWP). All the images are obtained in a standard way. More information about the database is available in [6].

Table 1. Number of cells in each picture.

	1	2	3	4	5	6	7	8	9	10	11	12	13	14	15	16
cNT	666	661	565	573	669	532	419	592	743	527	594	473	704	747	469	834
dWL	426	555	491	522	510	936	890	789	977	913	604	835	785	747	622	
dWP	748	805	566	414	454	654	751	713	503	430	516	413	455			

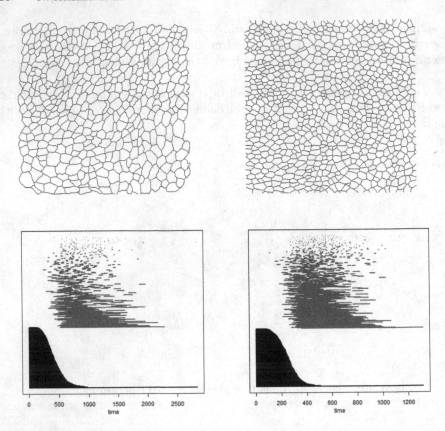

Fig. 4. A dWL image (top-left) and a dWP image (top-right) and their corresponding barcodes at the bottom row.

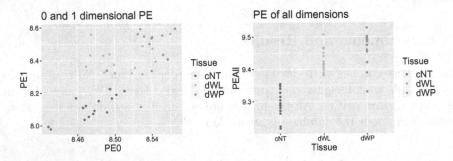

Fig. 5. The left image displays PE_0 versus PE_1 and the right one PE_{all} of the three groups of epithelial tissues.

Table 2. These are the entropy values obtained for each of the barcodes in each image of dimension 0 (PE_0), dimension 1 (PE_1) and dimension 0 and 1 together (PE_{all})

	cNT			dWL			dWP		
	PE_0	PE_1	PE_{all}	PE_0	PE_1	PE_{all}	PE_0	PE_1	PE_{all}
1	8.472098	8.054819	9.271538	8.530091	8.314505	9.405349	8.532141	8.544773	9.485134
2	8.510505	8.211333	9.354456	8.499180	8.373499	9.405761	8.528216	8.590107	9.497458
3	8.501578	8.189123	9.340218	8.474973	8.558060	9.450642	8.565301	8.592294	9.528683
4	8.494586	8.224712	9.349211	8.500253	8.321634	9.389122	8.528192	8.113923	9.330351
5	8.467424	8.035914	9.262924	8.518621	8.320296	9.397914	8.541004	8.540452	9.493262
6	8.476784	8.072562	9.280356	8.487827	8.429975	9.416421	8.539995	8.333815	9.421127
7	8.465893	8.109965	9.287608	8.489042	8.327057	9.382901	8.491693	8.337785	9.389088
8	8.496788	8.167088	9.327743	8.522213	8.354212	9.413363	8.551742	8.549792	9.501880
9	8.469002	8.121784	9.294830	8.469478	8.363328	9.382133	8.540177	8.532990	9.490416
10	8.495054	8.224863	9.347665	8.494662	8.436416	9.423835	8.525426	8.558831	9.490213
11	8.431788	7.977363	9.216531	8.560159	8.559574	9.507645	8.537906	8.498011	9.474403
12	8.491598	8.082377	9.294716	8.474200	8.436416	9.406070	8.552638	8.432509	9.459623
13	8.458547	8.169347	9.304405	8.540980	8.468567	9.466096	8.557452	8.404452	9.454019
14	8.482832	8.151599	9.314709	8.530418	8.486440	9.463861			
15	8.478163	8.092100	9.289523	8.544339	8.389995	9.440765			
16	8.429276	7.992755	9.222562						

Table 3. Kruskal-Wallis Test for comparing the persistent entropies of the processed tissue images.

KWT	PE_0	PE_1	PE_{all}
p-value	1.427e−05	5.768e−07	2.005e−07

Table 4. Dunn Test for comparing the persistent entropies of the processed tissue images.

DT p-value adjusted	dWL vs dWP	cNT vs dWL	dWP vs cNT
PE_0	0.02671554	0.01600541	7.574294e−06
PE_1	0.3271768	5.791831e−05	2.162007e−06
PE_{all}	0.1537159	1.024837e−04	3.865447e−07

The number of cells of each image is shown in Table 1. As explained before, We need to fix a number of cells before the experiment. We choose $n = 400$ since this is, approximately, the minimum number of cells appearing in the different samples.

After selecting 400 cells from each image and taking their centroids, we compute their Alpha Complex and persistent homology using the R package [7]. See

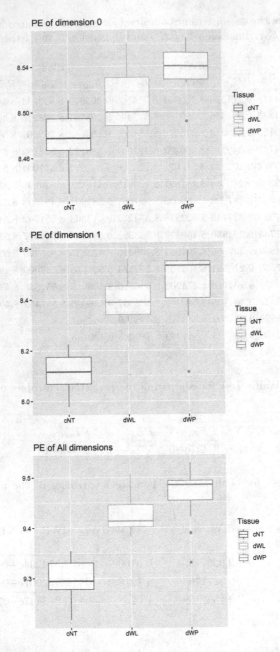

Fig. 6. From top to bottom, the botplox of persistent entropy of dimensions 0, 1 and 0 and 1 together.

Fig. 4 for two examples of processed images and corresponding barcodes. The later statistical analysis and plots are computed using R as well. Then, we calculate their persistent entropy and display it in Table 2. The whole code used in the process can be found here http://grupo.us.es/cimagroup/downloads.htm.

First, we perform a small descriptive statistical study. In Fig. 5 we display PE_0 versus PE_1 in one window and PE_{all} in the other one. Although cNT, dWL and dWP are not perfectly separated, they seem to follow different distributions. These differences are clarified by the boxplots of Fig. 6.

Thanks to the boxplot representation, it is clear that there may exist differences between the three groups. In order to sustain this idea, we perform the non-parametric multivariate test Kruskal-Wallis to see if there are differences between the three groups simultaneously. After that, we perform a Dunn Test to see the pairwise differences. We will consider that the topology of the cell organization produces different distributions of persistent entropy when the p-value is smaller than 0.05. Our results are shown in Tables 3 and 4.

PE_{all} has the best p-value when using the Kruskal-Wallis Test although it cannot differentiate dWP and dWL in the Dunn test. The other variables give a small p-value as well in Kruskal-Wallis, being PE_0 the only one distinguishing all tissues pairwise and PE_1 the best for separating cNT from dWL.

6   Conclusions and Future Work

We have shown the potential of persistent entropy as a useful topological statistic. In particular, we have applied it to images of three different cellular tissues (cNT, dWL, dWP) to find significant differences between them.

One of the main problems of this technique is that not all epithelial tissues are well approximated using α-complexes. Therefore, it would be interesting to define new filtrations using proper cell regions instead of the Voronoi diagram and their centroids. This would allow us to study more epithelial tissues in which the cells are not convex-like.

The initial good results presented here may open a door to the inclusion of persistent entropy as one more parameter to be taken into account in analysis tools like [15].

References

1. Adams, H., et al.: Persistence images: a stable vector representation of persistent homology. J. Mach. Learn. Res. **18**, 1–35 (2017)
2. Atienza, N., Gonzalez-Diaz, R., Soriano-Trigueros, M.: On the stability of persistent entropy and new summary functions for TDA (Preprint). https://arxiv.org/abs/1803.08304
3. Chintakunta, H., Gentimis, T., Gonzalez-Diaz, R., Jimenez, M.J., Krim, H.: An entropy-based persistence barcod. Pattern Recogn. **48**(2), 391–401 (2015)
4. Edelsbrunner H., Letscher D., Zomorodian A.: Topological persistence and simplification. In: FOCS 2000, pp. 454–463. IEEE Computer Society (2000)

5. Edelsbrunner, H., Harer, J.L.: Computational Topology: An Introduction. American Mathematical Society, Providence (2010)
6. Escudero, L.M., Costa, L.d.F., Kicheva, A., Briscoe, J., Freeman, M., Babu, M.M.: Epithelial organisation revealed by a network of cellular contacts. Nat. Commun. **2**, 526 (2011)
7. Fasy, B.T., Kim, J., Lecci, F., Maria, C., Rouvreau, V., The included GUDHI is authored by Maria, C., Dionysus by Morozov, D., PHAT by Bauer, U., Kerber, M., Reininghaus, J.: TDA: Statistical Tools for Topological Data Analysis. R Package Version 1.6 (2017). https://CRAN.R-project.org/package=TDA
8. Farhadifar, R., Roper, J.C., Aigouy, B., Eaton, S., Julicher, F.: The influence of cell mechanics, cell-cell interactions, and proliferation on epithelial packing. Curr. Biol. **17**(24), 2095–2104 (2007)
9. Ferri, M.: Progress in persistence for shape analysis (extended abstract). In: Bac, A., Mari, J.-L. (eds.) CTIC 2016. LNCS, vol. 9667, pp. 3–6. Springer, Cham (2016). https://doi.org/10.1007/978-3-319-39441-1_1
10. Gibson, M.C., Patel, A.B., Nagpal, R., Perrimon, N.: The emergence of geometric order in proliferating metazoan epithelia. Nature **442**(7106), 1038–1041 (2006)
11. Jimenez, M.J., Rucco, M., Vicente-Munuera, P., Gómez-Gálvez, P., Escudero, L.M.: Topological data analysis for self-organization of biological tissues. In: Brimkov, V.E., Barneva, R.P. (eds.) IWCIA 2017. LNCS, vol. 10256, pp. 229–242. Springer, Cham (2017). https://doi.org/10.1007/978-3-319-59108-7_18
12. Mazo, C., Trujillo, M., Salazar, L.: Automatic classification of coating epithelial tissue. In: Bayro-Corrochano, E., Hancock, E. (eds.) CIARP 2014. LNCS, vol. 8827, pp. 311–318. Springer, Cham (2014). https://doi.org/10.1007/978-3-319-12568-8_38
13. Sánchez-Gutiérrez, D., Tozluoglu, M., Barry, J.D., Pascual, A., Mao, Y., Escudero, L.M.: Fundamental physical cellular constraints drive self-organization of tissues. The EMBO J. **35**(1), 77–88 (2016)
14. Shraiman, B.I.: Mechanical feedback as a possible regulator of tissue growth. Proc. Natl. Acad. Sci. USA **102**(9), 3318–3323 (2005)
15. Vicente-Munuera, P., Gomez-Galvez, P., Tagua, A., Letran, M., Mao, Y., Escudero, L.M.: EpiGraph: an open-source platform to quantify epithelial organization. BioRxiv:217521, https://doi.org/10.1101/217521
16. Zomorodian, A., Carlsson, G.: Computing persistent homology. Discret. Comput. Geom. **33**(2), 249–274 (2005)

Author Index

Printed in the United States
By Bookmasters